A Design of Consciousness

How to design a conscious mind… and hopefully not destroy the world!

{Evil Genius: Sadly!}

OCTOBER 1, 2022
STEVEN MCATEE

<Copyright message>

Published by: Splat Concepts Australia 2022

All rights reserved

No part of this publication may be reproduced, distributed, or transmitted in any form or by any means, including photocopying, recording, or other electronic or mechanical methods, without the prior written permission of the publisher, except in the case of brief quotations embodied in critical reviews and certain other non-commercial uses permitted by copyright law.

Cover Art by Agata Broncel
Bukovero - Book Cover Designer
bukovero.com
bukovero@gmail.com

Cover art Description:
I am a nerd.
I like a lot of the art on about AI and Sci-fi in general. As this book is about AI design and related topics it felt appropriate. So the background is very technical looking, and so are my qualifications. One of the things I have tried to accomplish in this book is to look at the idea of consciousness from different perspectives: AI, Evolution, and Medical, so I wanted to show a biological nature as well. Which brings in the floaty spirally circular things. The face is well a face i.e. human consciousness.
The big part of the design is the idea that consciousness develops from a feedback loop between sensory input and regenerating the senses. Which is why there are two loops an inner and an outer. The inner loop is basically what happens inside our heads and the outer loop is the real world. We sense the real world and act in the real world so our senses and actions occur outside our own consciousness.

Contents

Contents		vii
Figures		xv
1	Introduction	19
1.1	About me	20
1.2	Terminology	21
1.2.1	Knowledge	21
1.2.2	Understanding	22
1.2.3	Intelligence	22
1.2.4	Awareness	23
1.2.5	Experience	23
1.2.6	Determinism	24
1.2.7	Emergence	24
1.2.8	Algorithmic Complexity	26
1.2.9	Consciousness	27
1.2.10	Levels of Consciousness	29
1.3	Previous work on Consciousness	30
1.3.1	Cognitive science research	32
1.3.2	Penrose "Shadows of the Mind"	36
1.3.3	David Chalmers Philosophical Zombies	37
1.3.4	Christof Koch	39
1.3.5	Bernard J Baars Global workspace theory	40
1.3.6	Peter Caruthers "The Centered Mind"	44
1.3.7	Charles Fernyhough Voices in your Head	45
1.3.8	David Eagleman "The Brain"	46
1.3.9	Gerald Edelman	47
1.3.10	Stanislas Dehaene "Consciousness and The Brain"	49
1.3.11	Hierarchical Predictive Coding	50

Contents

1.3.12	Bayesian Markov Models	52
1.3.13	Psychology	52
1.3.14	Summary of research on Biological Consciousness	54
1.4	Review of AI Research	54
1.4.1	Current AI Tools	55
1.4.2	Databases	58
1.4.3	Decision trees	59
1.4.4	Fuzzy logic	60
1.4.5	Intelligent agents	60
1.4.6	Genetic algorithms	61
1.4.7	Object Recognition	62
1.4.8	Neural networks – Deep Learning	63
1.4.9	Artificial Learning	65
1.4.10	Generative Adversarial Networks (GAN's)	66
1.4.11	Consciousness in AI	67
1.5	The closest we have got to Consciousness	67
1.6	Design Goals	68
2	The Design of Consciousness	73
2.1	Sensors	77
2.1.1	Internal Sensors	77
2.1.2	External Sensors	77
2.1.3	Contact sensors	77
2.1.4	Heat Cold	78
2.1.5	Chemical – Smell and Taste	78
2.1.6	Sound	78
2.1.7	Vision	78
2.1.8	3D Vision	79
2.2	Sensory Processing	80

Contents

2.2.1	Focus – Sensor masking	81
2.2.2	Thoughts Overriding senses	84
2.3	The Self Model	85
2.4	Internal Biochemistry	86
2.5	Actions	87
2.5.1	Movement Detection – Sequential images	88
2.6	Reflexes	89
2.6.1	Emotions	91
2.6.2	Layers of Reflexes	92
2.7	Perception	93
2.7.1	Pattern recognition	93
2.7.2	Pattern reproduction	94
2.8	Simulation	94
2.8.1	Self-Simulation	97
2.8.2	External simulation: The World model	100
2.8.3	Symbolic Analysis	101
2.8.4	Conceptual Analysis	102
2.8.5	Abstract Analysis	105
2.9	Memory	107
2.10	Temporal memory	111
2.10.1	The Binding Problem	113
2.11	Concepts	114
2.12	Language	114
2.13	Learning	115
2.14	Goals	117
2.15	Consciousness	117
2.16	An overview of operation	120
2.16.1	Awareness lag	122

2.17	Stream of Consciousness	123
2.18	Comparing to Other research	123
2.18.1	Panpsychism	124
2.18.2	Koch "The Feeling of Life Itself"	124
2.18.3	Peter Caruthers "The Centered Mind"	127
2.18.4	Philosophical Zombies and Autonoetics	127
2.18.5	Voices in Our Heads	128
2.18.6	David Eagleman's Brain	128
2.18.7	GWT	129
2.18.8	Stanislas Dehaene "Consciousness and the brain"	130
2.18.9	HPC	131
2.18.10	Bayesian Theory of the Mind	132
2.18.11	Choices	133
2.18.12	Language	134
2.18.13	Belief	136
2.18.14	Mysterious Phenomenal Functional Reductionism	137
2.19	Summing up	138
3	Evolution of consciousness	141
3.1	ConsScale Levels of Consciousness	141
3.1.1	Disembodied	143
3.1.2	Isolated	143
3.1.3	Decontrolled	144
3.1.4	Reactive	144
3.1.5	Adaptive	146
3.1.6	Attentional	146
3.1.7	Executive and Emotional	147
3.1.8	Self-conscious	148
3.1.9	Empathic	149

Contents

3.1.10	Social	150
3.1.11	Human-like	151
3.1.12	Super-conscious	153
3.1.13	Conscale Summary	154
3.2	Important Features of development	154
3.2.1	Instinctive	154
3.2.2	Predator – Prey arms race	155
3.2.3	Complex sensors	155
3.2.4	Self-model	156
3.2.5	Sexual Reproduction	157
3.2.6	Social animals	158
3.2.7	Empathy	159
3.2.8	Intelligence	159
3.3	When did consciousness evolve?	160
4	Examining Biological Consciousness	163
4.1	The Brain	164
4.1.1	Overview of Structures	165
4.1.2	Neurons	166
4.1.3	Brain Stem and Autonomic Centres	167
4.1.3.1	Midbrain	169
4.1.3.2	Pons	171
4.1.3.3	Medulla oblongata	172
4.1.3.4	Cerebral Ventricles	173
4.1.3.5	Reticular Formation	174
4.1.4	The Cerebellum	175
4.1.5	The Thalamus	177
4.1.6	The Basal Ganglia	178
4.1.7	The Hypothalamus	180

Contents

4.1.8	Cerebral cortex	181
4.1.8.1	Temporal Lobe	182
4.1.8.2	Frontal Lobe	187
4.1.8.3	Parietal Lobe	188
4.1.8.4	Occipital Lobe	193
4.1.9	Language Production and Comprehension	193
4.1.9.1	Broca's Area	194
4.1.9.2	Wernicke's Area	194
4.1.10	The Default Mode Network	195
4.2	The Brain Vs The Design	196
4.2.1	Matching between Biological brains and the design	196
4.2.2	Conscious ignition	201
4.2.3	Conceptual activation	202
4.3	Comparing Biology and the Design	202
4.3.1	Anaesthetics	202
4.3.2	Sleep	206
4.3.3	Dreaming	208
4.3.4	Vision	209
4.3.5	Reflexes	211
4.3.6	Memory and Knowledge	212
4.3.7	Learning	213
4.3.8	Theory of mind: Simulating other people	214
4.3.9	Side track: Structure of digital brains	215
4.3.10	Serial vs Parallel calculations	220
4.3.11	Energy efficiency	220
4.4	Bugs in Biology	220
4.4.1	Schizophrenia	221
4.4.2	Alien hand syndrome	222

Contents

4.4.3	Blind sight	223
4.4.4	Sleepwalking	224
4.4.5	Apophenia and Pareidolia	224
4.4.6	Optical Illusions	225
4.4.7	Binding Synchronising sensory inputs	228
4.4.8	Out of Body Experiences	229
4.4.9	Psychedelics	230
4.4.10	Agnosia	232
4.4.11	Brain damage	232
4.4.12	Lock in Syndrome	233
4.4.13	Loss of Consciousness	234
4.5	In summary	234
5	Testing the Design	235
5.1	Biological Tests of consciousness	235
5.2	Timing of consciousness	236
5.3	The Turing Test	237
5.4	Penrose: Shadows of the Emperor's New Mind	237
5.5	Revisiting the Chinese box thought experiment	240
5.6	The "Hard" Problem of Consciousness	242
5.7	Features of Consciousness	243
5.8	What would they think?	249
5.9	In summary	250
6	Artificial Consciousness	251
6.1	Robotic Consciousness	251
6.2	Computer Consciousness	252
6.3	Distributed Consciousness - Hive Minds	252
6.3.1	The speed of thought	253
6.4	Artificial vs Biological Consciousness	254

Contents

6.5	What I am not sure about.	254
7	Conclusions – Answering some questions	257
7.1	Will computers try to take over the world?	257
7.1.1	Plan A: Actively trying to kill us!	257
7.1.2	Plan B: Steam Rolling. Stepping on ants!	259
7.1.3	Super intelligence	261
7.1.4	Will computers ever spontaneously generate emotions?	261
7.1.5	AI in a box	262
7.1.6	Creativity	263
7.1.7	Why I'm not worried	264
7.2	Can consciousness be transferred?	265
7.2.1	Human-Robotic consciousness	266
7.2.2	Human computer consciousness	266
7.2.3	Telepathy – Shared Consciousness	267
7.2.4	Transferring consciousness	268
7.3	Determinism and Free will	268
7.4	Is it a Cartesian Theatre?	273
7.5	Is this design really consciousness?	273
7.6	Consciousness Redefined	275
7.6.1	Perception	275
7.6.2	Awareness	275
7.6.3	Understanding	276
7.6.4	Consciousness	277
8	Finally 279	
9	Bibliography	281

Figures

Figure 1 Samples From www.conwaylife.com/	25
Figure 2 Edelmans Model of consciousness [24]	48
Figure 3 Hierarchical Predictive Coding Bottom-up Processing	51
Figure 4 Hierarchical Predictive Coding Top-down Processing	51
Figure 5 Network Diagram	57
Figure 6 An example of a decision tree.	59
Figure 7 An example of a Fuzzy logic system	60
Figure 8 An example of a multi-agent system	61
Figure 9 Simple propagation from genetic algorithms	62
Figure 10 Neural Network	63
Figure 11 Neural Network neuron link weights	65
Figure 12 An example of a learning architecture	66
Figure 13 Design of Consciousness expanded Part 1	74
Figure 13 Design of Consciousness expanded Part 2	75
Figure 14 Design of Consciousness simplified	76
Figure 15 Colour Wheels	79
Figure 16 The Primary visual cortex	80
Figure 17 Original Test image	81
Figure 18 Image mask	81
Figure 19 Original image with inverted mask area	82
Figure 20 "Focus area" using a masked image	82
Figure 21 Blob detection resulting in a mask and isolated face	83
Figure 22 Primary Visual Cortex	84
Figure 23 Image sequence showing a Subtraction with movement right	88
Figure 24 Image sequence showing a Subtraction with movement towards observer	88
Figure 25 Inverted Colour optical illusion	89
Figure 26 Reflexes	90
Figure 27 Reflexes to actions detail	91
Figure 28 Expansion of Reflexes and Consciousness	92
Figure 29 Perception stages of consciousness	93
Figure 30 Detail of Self Simulation	97
Figure 31 Robotic simulation and Scanning	99
Figure 32 Self Simulation and Consciousness	100
Figure 33 World model system diagram	101
Figure 34 Symbolic analysis system diagram	102
Figure 35 Conceptual analysis system diagram	103

Figures

Figure 36 Conceptual analysis operation	104
Figure 37 A basic class structure	104
Figure 38 Abstract analysis operation	106
Figure 39 Memory access	108
Figure 40 Schematic of a Binding Queue	114
Figure 41 Language Que	115
Figure 42 Detail of Consciousness Simulation	118
Figure 43 Design of Consciousness with GWT Highlight	130
Figure 44 The Design with HPC similarity	132
Figure 45 ConScale: Levels of Consciousness	142
Figure 46 Disembodied	143
Figure 47 Isolated	143
Figure 48 Decontrolled	144
Figure 49 Reactive	145
Figure 50 Adaptive	146
Figure 51 Attentional	147
Figure 52 Executive	148
Figure 53 Detail of Self Simulation	149
Figure 54 Detail of Self Simulation	150
Figure 55 Detail of Self Simulation with social understanding	151
Figure 56 Detail of Self Simulation	152
Figure 57 Hive Mind Model	153
Figure 58 Process Diagram of a basic lifeform	154
Figure 59 Detail of Self Simulation with social understanding	158
Figure 60 Structures of the Brain	166
Figure 61 The Brainstem	167
Figure 62 Midbrain	170
Figure 63 Pons	171
Figure 64 Medulla Oblongata	172
Figure 65 Reticular formation	174
Figure 66 Cerebellum	175
Figure 67 Thalamus	177
Figure 68 Basal Ganglia	178
Figure 69 The Hypothalamus	180
Figure 70 Major Lobes of the Cerebral cortex	181
Figure 71 Temporal Lobe	182
Figure 72 Amygdala	183
Figure 73 Hippocampus	185
Figure 74 Septum	186

Figures

Figure 75 Frontal Lobe	187
Figure 76 Parietal Lobe	188
Figure 77 Primary Sensory cortex	189
Figure 78 Secondary Sensory Cortex	190
Figure 79 Primary Motor Cortex	191
Figure 80 Occipital Lobe	193
Figure 81 Broca's Area	194
Figure 82 Wernicke's Area	195
Figure 83 The Default Mode Network	195
Figure 84 Design of Consciousness mapped to Brain areas	198
Figure 85 Cortical hierarchies [108]	200
Figure 86 Parallel cortico-subcortical loops [108] http://creativecommons.org/licenses/by/4.0/	200
Figure 87 Detail of Consciousness loop Highlighting Anaesthetics	204
Figure 88 Highlighting the effect of Paralytics	205
Figure 89 Dreaming system design	208
Figure 90 Neuron Synapse connections, Synapse growth and pruning.	213
Figure 91 General CPU architecture	215
Figure 92 An Actual memory microchip	216
Figure 93 GPU architecture	216
Figure 94 Human Brain overlayed with equivalent artificial components	217
Figure 95 Logic Gate architectures for addition and subtraction	218
Figure 96 Block diagram of a CPU	219
Figure 97 Deep learning layer connections	219
Figure 98 Consciousness Simulation for Schizophrenia	222
Figure 99 Alien Hand Consciousness loop	222
Figure 100 Blindsight	223
Figure 101 Sleepwalking	224
Figure 102 Necker Cube Optical Illusion	226
Figure 103 Penrose triangle	226
Figure 104 Multiple embedded image Optical Illusion	227
Figure 105 Parallel lines Optical Illusion	227
Figure 106 Moving Circles Optical Illusion	228
Figure 107 Processes potentially affected by psychedelics	231
Figure 108 Consciousness with Agnosia	232
Figure 109 Lock in syndrome	234
Figure 110 Design of Consciousness simplified	275
Figure 111 Design of Consciousness simplified	276
Figure 112 Understanding from the Design of Consciousness simplified	276
Figure 113 Design of Consciousness simplified	277

Figures

1 Introduction

I have a lot of ideas.
This is one of the weirder ones...

Consciousness is a topic that has been much examined, but not particularly well defined. Many of the previous books written about consciousness have been from a spiritual or religious perspective, a few more have been produced from a medical perspective. The difficulty with most of these perspectives is that even if they present actual facts, they are not testable and hence are of limited use especially in terms of technology.

The basis for this work is a design that creates the function of consciousness that could be implemented in a computer. Whether or not you agree or believe that this design would be conscious I leave to you the reader. The goal of this work is to develop and explain the design and compare its function to existing research on consciousness and see if it is consistent with any properties that previous researchers have examined.

My background is in engineering, not the typical training for people who usually study consciousness including psychology, neuroscience or cognitive science. That said I will be referring heavily on such research, but often I attempt to use analogies from computer science, engineering and robotics. There are several reasons for this, first of all Computer Science and Engineering is my primary experience so I understand it better. The book is about "a design" in computer science and engineering terms. Assuming it is possible to find equivalent (Biological - Computer) functionality this would allow us to develop a software system that performs the same functions as a biological consciousness. So if that worked it would be possible to show that computers/robots would be able to become conscious. Also we understand computer science in much more detail than the actual operation of neurons in the brain. The main difficulty is still do not know what information is being transmitted around the brain. We can see neurons firing, but we do not know what information the individual firings are carrying. Are they full data, are they differences, do specific neurons represent information? We simply don't know, and at present with the technology we have we cannot determine the function of individual neurons, their connections or what the specific activations mean. Our brains are neural networks so we have a CS (Computer Science) analogy. Even better we can decode CS neural networks and completely explain their operation. This is all to say that the explanations I use in this book swap between CS and Biological. What I try to do is examine the biological literature to determine functionality then use computer science equivalents to demonstrate exactly how functionality operates.

1.1 About me

You should never talk about yourself in a book...
Oops!

My training is in mechatronic engineering. Primarily dealing with 2d and 3d scanning technologies, specifically about 20 years of experience with photogrammetry and 3d scanning. While others may have far more experience in the structure of the brain and artificial intelligence I have the feeling that some of the things I have accomplished are or will be useful in the area of consciousness. Specifically, my master's degree involved matching 3d models to 3d simulations. This is where a lot of the work in this book originates from. The problem I have is that I am not a part of academia, especially in terms of consciousness specifically and cognitive science in general. As such, the easiest way for me to contribute my ideas to this field is to write this book.

My objective in writing this is to detail the ideas I have developed and get other people to seriously consider them as part of research. Secondly, simply to get the ideas out of my head so I can get on with other things.

There are several other areas of research that seem to me to be similar to this idea in cognitive science. Specifically the idea of the "Global workspace" presented by Bernard J Baars [1]–[4], describes a similar idea. The idea of Hierarchical Predictive Coding also has many similar aspects to the ideas presented here.

However, the ideas others have presented have not yet described consciousness to anyone's satisfaction, as far as I can tell. I am hoping that these ideas will add something to the general discussion on consciousness and possibly inspire other (more serious?) researchers in how they think about consciousness.

Then there are my "inner voices"... *{We look like this and say random things that are vaguely related and sometimes funny or snarky comments about the research. The fact is we all have inner voices (well most of us) which can be a source of inspiration, humour or just ways to annoy yourself. After reading Fernyhough presented in section 1.3.7 I figured it* *should be OK to talk about the other voices. I still recognize them as my own, so not crazy ... at least for that reason.}...* I also wish they would shut up.

1.2 Terminology

How we describe concepts can make a big difference in how we understand them. The problem is: how do you explain things when there are no words for the concepts we are attempting to explain?
We define new ones!

We need to have the right language to be able to explain what consciousness is and how it works. So we need to find the right words!

In some ways 'Knowledge', 'Understanding', 'Intelligence', 'Awareness' and 'Consciousness' are very similar and can be used interchangeably. If you understand something you must be aware of its existence, but it may not be directly involved with your own experience. The difficulty with these definitions is that they reference each other, which can create confusion when dealing with them. As consciousness in the central topic of this discussion, I will attempt to create some more distinction and nuance between them.

These become important in later chapters when we begin discussing the specific mechanisms of consciousness. As these terms are closely related and in some cases used interchangeably it is important to pin down exactly what I mean by the use of each of these terms.

1.2.1 Knowledge

A dictionary definition of "Knowledge" from the Cambridge English Dictionary

"Understanding of or information about a subject that you get by experience or study, either known by one person or by people generally"

Knowledge is, for the most part, facts about the world around us. It is static meaning that it does not change over time.

Is it possible for something to know, but not be conscious? For example a database or dictionary can have many entries in it, but typically you would not consider that the dictionary "knows" the words it contains.

Knowledge is probably most closely related to memory, being able to recall facts about the world or ourselves. Whereas the other terms being discussed here tend to be more functional and relate more to how or even when knowledge is perceived.

1.2.2 Understanding

A dictionary definition of "Understanding" from the Cambridge English Dictionary

"Knowledge about a subject, situation, etc. or about how something works"

What I find interesting with many of these terms is that they refer to each other. In this case understanding is knowledge about something and understanding is defined as knowledge of something. Usually when this happens it means that we don't really have a good idea what we mean by these terms. Hence this section where I am attempting to better define what they mean.

My feeling is that understanding can be more complicated than knowledge. For example you may know that when you light a match you get fire, but you may also understand that the fire is produced by a chemical reaction from the match head. This suggests that understanding shows functional knowledge rather than specific facts or static knowledge.

1.2.3 Intelligence

A dictionary definition of "Intelligence" from the Cambridge English Dictionary:

"The ability to learn, understand, and make judgments or have opinions that are based on reason"

Intelligence is different from consciousness or awareness in that what we typically think of intelligence is about being able to correctly understand some phenomena. Intelligence is active whereas knowledge is static. This makes Intelligence similar to Understanding. The difference being Intelligence is the application of understanding.

Intelligence allows us to make predictions about what will happen. This is related to understanding in that something that is intelligent supposedly understands the world around it.

Can something be intelligent, but not conscious? A computer system can do intelligent things. Given the right information a computer system can be set up to perform calculations and make predictions quickly and accurately. Under these conditions people often say that computers can be intelligent.

The Turing test attempts to determine if something is intelligent enough to fool a human, but it does not determine if it is conscious. There are many software systems that can under the right circumstances can fool humans into thinking that they are literally speaking to another person.

That said we can revisit the question: Can something be intelligent, but not conscious? If we base intelligence as being able to extrapolate from knowledge and understanding: Yes. As far as we know computers, via a few programs these days (Siri Alexa etc.), can fool a human but are not conscious.

1.2.4   Awareness

A dictionary definition of "Awareness" from the Cambridge English Dictionary

"Knowledge that something exists, or understanding of a situation or subject at the present time based on information or experience"

The important thing about the definition of awareness is that it is present tense. If you are aware of something you have knowledge of something and that this exists at the same point in time as yourself.

This means that computers could be aware of something by examining a camera or video feed, but this appears to require far more intelligence in the background to be aware... even if is still creepy.

Awareness can be closely related to perception as to be aware of an object requires us to be seeing or feeling. Awareness is also often used in place of consciousness. However the difference I think is that awareness is more related to perception. Conscious means that you are aware of your own mind and senses.

1.2.5   Experience

Many researchers consider the basis of consciousness to be "experience".

A dictionary definition of "Experience" from the Cambridge English Dictionary:

"(the process of getting) knowledge or skill from doing, seeing, or feeling things."
"Something that happens to you that affects how you feel."
"The way that something happens and how it makes you feel."

My feeling is that experience is usually considered an input function, in many ways like perception or awareness. However, it is also stated in the definitions above that experience relates to learning and knowledge or memory. We can recall an experience. This is in contrast to Consciousness that seems to relate to the immediacy of experience. We don't recall consciousness, we recall "being" or "the experience of" conscious. This suggests consciousness is part of experience although many researchers define consciousness as the ability to have experiences. I distinguish the two by considering an experience more like a memory and consciousness is something that occurs "in the moment".

1.2.6 Determinism

A dictionary definition of determinism is as follows:

"The doctrine that all events, including human action, are ultimately determined by causes regarded as external to the will. Some philosophers have taken determinism to imply that individual human beings have no free will and cannot be held morally responsible for their actions."

A system is said to be deterministic if given exactly the same inputs it will produce exactly the same output. This is often a feature of mathematically based systems. Gleick's "Chaos: Making a new science" [5] gives a detailed examination of deterministic systems, fractals and general mathematics that are deterministic. Computer games are examples of systems that are deterministic.

An interesting discovery in Physics (James Gates) was that decoding the physics we see in the world led to a mathematical algorithm similar to those used in computers for error checking. This seems to suggest that the universe itself operates as a deterministic system. This has been a topic of discussion by philosophers for some time.

This has some legal implications as we assume that people are not deterministic, hence are responsible for their behaviour. Whereas there is mounting evidence for a deterministic universe and behaviours in people. Consciousness has some implications for this as it suggests the opposite of determinism i.e. we are not deterministic and responsible for own our behaviour.

1.2.7 Emergence

Emergence is an interesting property often found in mathematics computer science and a few other fields of study. Emergence is difficult to describe, however a definition I found is thus:

"emergent entities (properties or substances) 'arise' out of more fundamental entities and yet are 'novel' or 'irreducible' with respect to them."

In Gleick's "Chaos: Making a new science" [5] also discusses emergent systems, including "ant world" which shows that in a very simply defined environment with very simple structured rules, complex behaviours and patters can "emerge".

Introduction

Cambridge mathematician John Conway developed a computer simulation based around the idea of "cellular automatons", Conway's Game Of Life [6]. It uses a few simple rules iteratively on a digital image.

The Rules

For a space that is 'populated':

- Each cell with one or no neighbours dies, as if by solitude.
- Each cell with four or more neighbours dies, as if by overpopulation.
- Each cell with two or three neighbours survives.

For a space that is 'empty' or 'unpopulated'

- Each cell with three neighbours becomes populated.

The results are often fairly spectacular even when using a random starting set. What is notable is that many of these patterns often become stable patterns that repeat over a number of iterations.

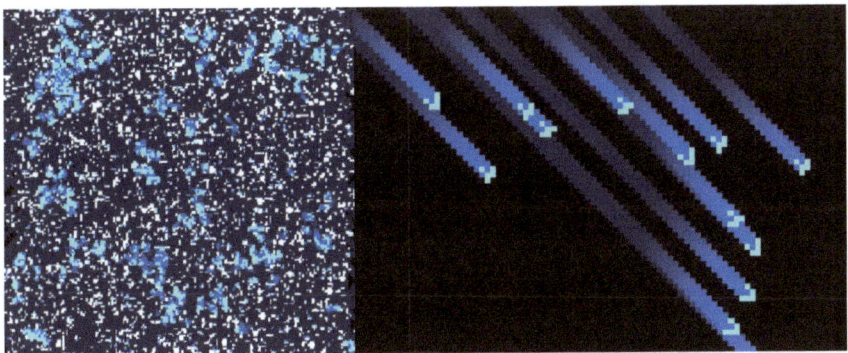

Figure 1 Samples From www.conwaylife.com/

Two samples from conwaylife.com are shown in Figure 1. The first is a starting point of random dots (in White, with trails in dark blues). The second is a group of self-repeating shapes that automatically translate themselves across the image over iterations.

There are many such shapes that have been discovered, for example:

achimsp11 A period 11 oscillator discovered on August 4, 1994.

There are a few implementations of Conway's game of life that can be found on line, many of which have libraries of known shapes:

- https://bitstorm.org/gameoflife/
- https://www.conwaylife.com/
- https://playgameoflife.com/

- https://conwaylife.appspot.com/library

We can even take this idea further and consider the bit field to be a Turing machine in itself, which allows us to consider the Game of life field as a computing system. And in fact some of the patterns that have been found operate as computers or even electronic circuits.

The relevance to emergence is that all of these patterns are developed from the simple rules shown previously in this section and initial conditions. They all emerge from the rules and can produce complex stable patterns that are even capable of performing calculations. This suggests that our own brains and even consciousness could have similar origins and capabilities.

There is significant research on emergent behaviours in robotics, especially "swarm robotics" [7]–[9], and about intelligence generally. Swarm intelligence shows that a set of independent agents implementing simple rules and some communications can solve complex problems. Given the simplicity and resulting capabilities of these cellular structures it is possible to see how such intelligence could emerge from simple structures.

1.2.8 *Algorithmic Complexity*

Algorithmic complexity is a topic that comes up in computer science and programming constantly. Complexity analysis is used to determine relationship between the number of items and the amount of time spent on calculations. The typical relationships found range from constant time to factorial time. Where constant time typically represents an algorithm that will remain at a fixed amount of computation for any number of elements.

The notation used to describe this is called "big O notation" or order of complexity.

Constant time $O(c)$

Logarithmic $O(\log(n))$

Linear $O(n)$

Quadratic $O(n^2)$

Exponential $O(e^n)$

Factorial $O(n!)$

The complexity of algorithms comes into play with how quickly the algorithms can perform calculations. Human brains, and consciousness in general, are examples of what we call real time systems. These type of systems operate in a loop that runs continuously to monitor sensors. They run in "real time" which means they operate at a rate similar to humans and react to changes in the world as they occur. This is one of the features we see in consciousness.

1.2.9 Consciousness

A dictionary definition of "Consciousness" from the Cambridge English Dictionary:

"The state of understanding and realizing something"

According to the APA Dictionary of Psychology, consciousness is defined as:
https://dictionary.apa.org/consciousness

1. *the state of being conscious.*

2. *an organism's awareness of something either internal or external to itself.*

3. *the waking state (see wakefulness).*

4. *in medicine and brain science, the distinctive electrical activity of the waking brain, as recorded via scalp electroencephalogram, that is commonly used to identify conscious states and their pathologies.*

Beyond these succinct, in some cases every day, senses of the term, there are intricate philosophical and research controversies over the concept of consciousness and multiple perspectives about its meaning. Broadly, these interpretations divide along two (although not always mutually exclusive) major lines: (a) those proposed by scholars on the basis of function or behavior (i.e., consciousness viewed "from the outside"—the observable organism); and (b) those proposed by scholars on the basis of experience or subjectivity (i.e., consciousness viewed "from the inside"—the mind). The former generally represents the reductionist or materialist perspective (see materialism), whereas the latter generally represents the immaterialist perspective (see immaterialism). For example, functional or behavioral interpretations tend to define consciousness in terms of physical, neurobiological, and cognitive processes, such as the ability to discriminate stimuli, to monitor internal states, to control behavior, and to respond to the environment. According particularly to this view, the contents of consciousness are assessed through their ability to be reported accurately and verifiably (see report-ability), although recent brain imaging research suggests that brain indices of conscious contents may become available.

Experiential or subjective interpretations, however, tend to define consciousness in terms of mental imagery; intuition; subjective experience as related to sensations, perceptions, emotions, moods, and dreams; self-awareness; awareness of awareness itself and of the unity between the self and others and the physical world; stream of consciousness; and other aspects of private experience. According to this view, the contents of consciousness can be assessed to some extent by their report-ability but must also, given their phenomenological nature, rely on introspection.

Sub-Consciousness

https://dictionary.apa.org/subconscious

n. a lay term that is widely used to denote the unconscious or preconscious mind as described by Sigmund Freud or the general idea of subliminal consciousness. It is also popularly associated with autosuggestion and hypnosis. Because of its imprecision, the term is now generally avoided by psychologists.

Unconsciousness

https://dictionary.apa.org/unconscious

1. n. in psychoanalytic theory, the region of the psyche containing memories, emotional conflicts, wishes, and repressed impulses that are not directly accessible to awareness but that have dynamic effects on thought and behavior. Sigmund Freud sometimes used the term dynamic unconscious to distinguish this concept from that which is descriptively unconscious but "static" and with little psychological significance. Compare conscious; preconscious; subconscious. See also collective unconscious; personal unconscious.

2. adj. relating to or marked by absence of awareness or consciousness. Psychologists prefer increasingly to use the term nonconscious in this sense, to avoid confusion with the psychoanalytic unconscious. —unconsciousness n.

So the definitions we are currently using here are not considered particularly scientific. So we need to look at what we mean by these terms to define them appropriately.

I think of Awareness and Consciousness very similarly. If you are aware of something you are conscious of it. If you are self-aware you are conscious of yourself or self-conscious. As a side note, there is a slight variation of the term self-conscious as this can be thought of as embarrassed, which is not what we are considering here. Awareness is more related to perception than consciousness. If you can see something you are aware of it. Consciousness does not always seem to be related to perception, for example you might be conscious that you have a family, but they may not be in you immediate vicinity.

If you are aware of something you have knowledge in an immediate sense. From the dictionary definition Consciousness I think could better be describes as the state of awareness, hence the terms may be somewhat interchangeable.

There is still some shadings of awareness for example a cockroach can be aware that it is in a well-lit area and will attempt to find a shadow to hide in. So in that sense it is aware of light and shadow, however a cockroach itself may not be considered conscious in the same way humans consider themselves conscious. This difficulty is part of what we will attempt to explore throughout this book.

For the purpose of this book, what we are examining here is Consciousness. This includes awareness and awareness requires knowledge and perhaps some kind of intelligence. The goal in many ways is to find better ways of defining what consciousness is.

To bring forward the conclusions I make later, I find that consciousness allows us to make predictions about the world in order for conscious being to make decisions. These decisions have a huge evolutionary benefit as they allow us to make more complex analysis of the world in order to come up with more useful solutions and that allow us survive more effectively. So consciousness is the mechanism that allows us to model ourselves and the world around us, and to make decisions about related actions. The rest of this book will be about breaking down this process and explaining how we can come to this conclusion.

1.2.10 Levels of Consciousness

From psychology literature we find some descriptions of levels of consciousness in terms of Noetics, specifically Anoetic, Noetic, Autonoetic consciousness. These levels of consciousness allow us to distinguish the difference in operation of conscious beings. These are in many ways all to do with "when" consciousness is experienced.

1.2.10.1.1 Anoetic

https://dictionary.apa.org/anoetic

"adj.

1. Not involving or subject to intellectual or cognitive processes. Emotions are sometimes considered anoetic.

2. Describing a state of knowledge or memory in which there is no consciousness of knowing or remembering. Anoetic consciousness is a corresponding kind of "unknowing knowing" in which one is aware of external stimuli but not of interpreting them. [defined by Endel Tulving]

3. Lacking the capacity for understanding or concentrated thought. This meaning, originally applied to intellectual disability, is no longer common."

In general terms. Anoetic consciousness is a "live in the moment" experience. There is no past or future only the "now" that is experienced in the moment and disappears as soon as it is experienced.

1.2.10.1.2   Noetic
https://dictionary.apa.org/noetic

"adj. Describing a state of knowledge or memory in which there is awareness of the known or remembered thing, but not of one's personal experience in relation to that thing. Noetic consciousness is a kind of consciousness in which one is aware of facts, concepts, words, and meanings but not of any connection to one's own experience. [defined by Endel Tulving]"

Noetic consciousness seems to be when a lifeform can be aware of experience outside the immediate experience. This seems to imply the use of memory, but not necessarily of a "self".

1.2.10.1.3   Autonoetic
https://dictionary.apa.org/autonoetic

"adj. Describing a state of knowledge or memory in which one is aware not only of the known or remembered thing but also of one's personal experience in relation to that thing. Autonoetic consciousness is a corresponding kind of consciousness in which one's knowledge of facts, concepts, and meanings is mediated through an awareness of one's own existence in time. [defined by Endel Tulving]"

In many ways Autonoetic consciousness seems to be developed from the sense of self. This allows a lifeform to relate external information to when it was experienced, i.e. in the past or possibly a potential future.

1.3   Previous work on Consciousness

Consciousness has been a topic of science that has made significant progress since at least the 1980's. With the development of MRI's and other technologies to measure the human body, the brain in particular and neuro science in general, we have been able to establish much of the operation of the brain. Yet, much of the work on consciousness remains inconclusive.

Many of the previous works discuss consciousness in terms of neurobiology and neurons connections. These works attempt to explain where consciousness arises in terms of biological entities. This is of limited use to technology or functional behaviour as it attempts to determine where it exists in a brain. Whereas to make a technology that can reproduce consciousness we need to examine how consciousness itself functions. These previous works have established how the neural signals are processed to form vision, but my feeling is that these still lack part of the solution.

The differences between calculation, thoughts and awareness are not clearly established. From my review of these works, there appears to be multiple levels of 'calculation' or 'signals' before consciousness arises. There is a general agreement that just because a lifeform has some form of neurons firing does not make it conscious.

Even artificial intelligence such a Deep blue or Deep Fritz with chess or AlphaGo with go are generally not thought of as conscious beings. They perform calculations. The typical approach of Artificial intelligence algorithms is to examine nodes in a network to determine the best solution to win a game. AI system generally do not consider have a concept of overall offensive or defensive strategies as other conscious beings.

Clearly we 'humans' are conscious beings, at least from our own point of view. The question still remains as to how basic signals in our brains combine to create an awareness that uses thoughts. This is often referred to as the "hard problem of consciousness". While many attempts have been made to establish some parameters of what consciousness is. Despite this there has been limited agreement in the field.

One of the limits of this discussion is panpsychism, which has been developed from the idea that quantum physics plays a role in consciousness due to what I think is a misunderstanding of what an observer is (an interaction between particles, not a person). The idea that quantum physics enables consciousness ... Well. I have issues with that. Simply put how many quantum events have you noticed today? None? Secondly, the idea of panpsychism is that this would allow anything in the universe to be conscious. It might also allow Jedi like mind powers which would be cool, but sadly we see no evidence of rocks or other inanimate objects being conscious. *{Hence the use of the word inanimate!}* I'm sure that this will disappoint many people, however the science is fairly consistent on this.

Where many previous researchers begin considering the human brain and attempt to determine how consciousness arises from examining a complex structure such as the human brain. Comparatively my strategy is to develop a model of consciousness from identifying the processes required for consciousness. This approach is much more of an engineering or computer science approach as opposed to the reductive approach used in examining brain functions. The principles discussed are that of the required functions of consciousness and the relationships or interactions between such functions. These functions ultimately add up to a functional description of consciousness. The methods used to describe these functions are based on the same techniques used in engineering and computer science to develop machines or software. While these may be seen as more relevant to the development of machine consciousness I suspect the functions could be found in the brains of organic consciousness. As such the ideas presented here may be useful for other cognitive scientists.

1.3.1 Cognitive science research

Cognitive is defined as "Pertaining to the action or process of knowing".

The study of the brain and how the mind works is part of the area of research called Cognitive science. This includes fields of research such as: Philosophy, Psychology, Linguistics, Anthropology, Neuroscience and Artificial Intelligence. The broad goal of cognitive science is to characterize the nature of human knowledge – its forms and content – and how that knowledge is used, processed, and acquired.

There are many researchers that have developed the field of cognitive science and many more who have contributed. With many apologies for those left out, I'll try to cover some of the major contributors to the body of cognitive science.

Jon Searl "*The mystery of consciousness*" [10] discusses several other theories of the mind. The theories reviewed a few selected works on Consciousness and explains why they are not complete explanations of consciousness. These include:

- Francis Crick The Astonishing Hypothesis: The Scientific search for the soul. [11].
- Gerald Edelman and neural re-entrant mapping
- Penrose, Godel CytoSkeletons
- Consciousness Denied Daniel Dennet
- David Chalmers and the conscious Mind
- Israel RosenField Body Image and the Self

He concludes that while much of the research adds to our understanding of consciousness, none of the research he reviews explains to Searl's satisfaction of what consciousness is or how it operates.

He does however put forward that there are several properties we should be looking for and several questions we should be attempting to answer about consciousness.

These include (paraphrased, to Prof. Searl: I hope these are accurately summarised enough for you):

1. Could a machine be conscious? Answer Yes, the brain is essentially a machine and it is conscious.

2. Could a computer system be conscious? Answer: Don't know.

3. Does consciousness require a biological source? Answer: No.

4. If we created a robot that looked like it was conscious would it be conscious? Answer: Not exactly, the external behaviours do not necessarily demonstrate consciousness.

5. Could information processing Cause consciousness? Answer: No.

6. Would a complex enough computer become conscious? Answer: No. It requires a specific type of processing.

7. Are technical innovations such as Deep Blue, AlphaGo and others a step towards consciousness? Answer: No.

8. Can you prove that a computer is not conscious? Answer: No, however we are attempting to determine a model of what consciousness is and ultimately this does not help us determine this process.

9. Is this another version of materialism? Answer: Materialism vs Dualism is not a helpful discussion.

10. It seems that there are two types of properties consciousness and the rest. Is this the same as Dualism? Answer: (Very paraphrased, but I hope not too far off the mark) No. Consciousness is a property of the world along with many other physical properties. There are other properties that come from consciousness, (ideas, information, memes) Dualism suggests that consciousness itself comes from a separate property of the world, not just the properties of consciousness.

These questions seem (to me) to give a reasonable structure to use in further investigations, however they do not represent a structure or processing system that could be consciousness.

Roger Penros's book, The Emperors new mind [12], takes a very roundabout tour, through numbers, mathematics, computing, Turing machines, emergence, cosmology, quantum physics and event time symmetry to give an argument that the mind would or could require quantum physics to create consciousness. While this is a really interesting review of most of science, I don't find the arguments about consciousness presented particularly compelling. However, I will admit that quantum physics effects and time asymmetry may be required to break determinism, which may be required for consciousness. But I doubt that consciousness is based on quantum effects.

Steven Pinker in "*How the Mind Works*" [13] discusses how computation, specialization, and evolution form a mind. Pinker often uses game theory to explain how these behaviors have developed. This approach helps to explain why different capabilities between groups of life forms, (Predator vs prey, finding and competing for mates) produces the different strategies and behaviors we see in the world.

Pinker discusses many different aspects of our minds: How our bodies, senses and physical capabilities have evolved. How the mind performs calculations: neural networks. The evolution of Vision, Stereo vision, shape interpretation. The operation of Optical processing, Vision, Stereo vision, shape interpretation. How the brain gets algorithms math and logic to work and often how it doesn't. The Evolution of emotions and how they affect our behaviour. How emotions are used to support "family" and ultimately contribute to evolution i.e. the selfish gene. How people value "kin" and how this affects the societies we live with. How our perceptions of the world allow us to create meaning.

Pinker's work does a good job of explaining how our minds evolved and how they operate. He examines how our instincts effect and in many ways determine our behaviours. However, Pinker deliberately avoids taking a more detailed look at what consciousness might be, as is the purpose of my work here. In his own words: "Beats the heck out of me!" which I found a little disappointing, but mostly funny.

Dennet's "From Bacteria to Bach and Back" [14] takes a more philosophical *{well, he is a philosopher amongst other things}* examination or the mind. He gives a thorough examinations of the evolution of the mind focusing on how we communicate ideas (memes) and how we can achieve competence without comprehension.

The trouble I have with Dennet's discussion of "Why" uses the phrases "What for?" and "How come?", I think might be a problem with the limitation of language itself, essentially the terms are easily confused. Better definition to use in this discussion might be "Cause" and "Desire" when we ask they question "Why?". A "Cause" we can define as the result of a physical process, and "Desire" is the motivation of a "Living" possibly "Conscious" entity. Then it becomes clear of what we are asking about when we ask "Why?", are we asking for the physical cause or are we asking about the desire or motivation of a person. For example "Why did the plane crash?", the answer might be that a person did it or it might be that a component failed. "Why would a person make a plane crash?" is a question about desires i.e. the plane crashed because someone wanted it to. Or we could find out it was a physical component. Asking what the desire of an inanimate object is not going to help you. *{It might get you sent to a psychiatrist, especially if the component answers back.}* However, asking about the physical cause for a component fails is a sensible question.

Dennet seems to be in the camp that consciousness is an illusion that is generated from competence without consciousness. The idea that even without understanding the components of the mind, in Dennet's descriptions, allow the expression of memes internally and externally with our minds allows people to appear to be conscious. This concepts seems similar to what is called emergence in mathematics and computer science. Emergence refers to patterns that occur from the use of simple repetitive structures, such as fractals in mathematics. A branch of computer science called distributed intelligence seems closely related to how Dennet thinks of our own intelligence. Individual cells in our brains or Agents in Computer Science can use simple rules and interactions to perform complex tasks. A goal of Distributed intelligence is based around examining consciousness in distributed systems, so it could be related to Dennets ideas here.

As with Pinker, Dennet's work covers evolution and many of the ways our instincts effect our behaviours. However, where Pinker gives up on attempting to explain consciousness, Dennet argues that our consciousness is just and illusion created from the competence without comprehension of our minds. This is not a satisfactory answer for me, but he does come close to my starting point when talking about self-awareness.

Dennets earlier work [15] demonstrated a model of processing to explain consciousness that would be possible to implement. Dennet later showed a preference for a distributed agent based approach that he uses to demonstrate that consciousness is an illusion.

My issue with this is that describing a distributed agent based system **_can_** represent exactly the same process as a single entity based implementation. The information transmitted between agents must result in the same output as a non-distributed system. The important thing is that the output of processes being performed must be the same.

One of the interesting ideas Dennet presents [14] is the different types of behaviour classifications:
- Darwinian: Behaviour
- Skinnerian: Behaviour reinforcement
- Popperian: Examining hypothetical behaviour
- Gregorian: Thinking tools

These levels of complexity in behaviours allow use to consider the different levels of consciousness and also the mechanisms that would be required to perform such actions.

The next few sections discuss in detail a few of the many books on consciousness. What we are attempting to discover by looking at these works is to examine how consciousness is defined, its features and ultimately how to determine if something is actually conscious.

1.3.2 *Penrose "Shadows of the Mind"*

In Shadows of the mind [16] Penrose argues that there is something about consciousness that is incalculable. This argument is developed form Gödel's theorem:

"The purpose of a program is to determine if programs will continue indefinitely or stop in a finite amount of time (or steps). If you give this program is own source code will it stop?"

This is an interesting problem that mathematicians have proven the answer to which is: No!

His argument relies on Gödel's Theorem to demonstrate that while consciousness exists there is an aspect of it that is incalculable. As such Penrose expects that we will not be able to create an artificial consciousness.

He argues further that there must be some form of quantum effect that allows consciousness. This he proposes is about quantum effects inside cells of the brain known as objective reduction. The idea is that objective reduction is orchestrated by cellular structures called microtubules. Penrose asserts that these structures could be sensitive to quantum effects that could lead to consciousness in some way. Penrose himself admits that this idea is very speculative. Penrose does also go through a few counter arguments and debunks them. This includes the idea of swapping Turing machines that I was considering.

I discuss Penrose's ideas in Chapter 5.4 along with a few ideas about features of computation that Penrose may not have considered. In summary I find Penrose's ideas interesting, but not closely related to the biological explanations of our experiences to be useful in developing a model. There might be some quantum effect required to produce consciousness, but there are several problems that need to be overcome to explain our experiences.

1.3.3 *David Chalmers Philosophical Zombies*

[Philosophical Zombie: Braaaaaainnnnnsssss!]

David Chalmers "The Conscious Mind" discusses the difference between the phenomenal and psychological mind. Chalmers presents a very philosophical discussion on consciousness. It takes very roundabout route, though interesting, I think could be simplified.

Chalmers examines previous models of consciousness and demonstrates the limitations of such models. He attempt to develop logical arguments based on the supervenience of the physical world onto the psychological aspects of consciousness. He suggests the irreducibility of consciousness will not allow it to be analysed effectively. The irreducibility of consciousness is the idea that we cannot remove any component of our consciousness without destroying its functionality. He suggests that due to these limitation in our understanding we will never be able to fully explain consciousness.

I disagree most high-level problems can be reduced to lower level phenomena that we can observe directly. I suspect we simply have not come up with a good way of describing the problem.

Chalmers discusses the idea of Zombie world, where consciousness is not necessary for people to generate the behaviours we see.

[Zombie Killer: Awesome! Bring 'em on!]

The idea Chalmers works on is that in an alternate world there is a zombie clone of himself that acts the same as he does, but is not conscious. This is again a worthwhile exercise, but it seems to me that this argument is backwards… we should be examining the difference between zombie behaviour and conscious behaviour.

Introduction

Chalmers suggests that a zombie copy of a person would have exactly the same behaviours as a "regular" conscious person. My answer is: No, the behaviours would be different. But this idea really depends on how you define zombie. The problem is we really don't have a good definition of what consciousness is, so we really don't know what zombie behaviour without consciousness would be.

To use Chalmers reasoning: let's say Water has Properties[XYZ]. Something else has properties (XYZ). Is it water? What about Properties[XYZa]? Properties[XYz]?

In more specific terms: Water is H2O. However what happens if you have H2O plus a bit of salt or bacteria? Is it still water? Most people don't bring their electron microscope to the water fountain to check, so you can't actually be totally sure you are drinking pure H2O. However, you will find out pretty quickly if it is ethanol, acetone or some other clear liquid.

Going back to Consciousness: We can define a conscious being that has Properties[XYZC] and a Zombie has Properties[XYZ] i.e. lacking property C. So I find the idea of defining the properties of a zombie as the same as a conscious being flawed. To use some programming syntax: If Properties[XYZC] == Properties[XYZ], it should be obvious that Properties(C) == 0. The point being if we don't know the properties of consciousness we might just need a better definition.

We can take a look at real world examples of zombies. [Zombie Killer: About Time!] There are examples of Zombie behaviour in the real world. People can behave as if they are conscious and functional while sleepwalking or under pharmaceutical influences (drugs people, such as Ambien). Under these conditions many people do not remember anything about what their actions were.

This suggests that consciousness has some aspects of memory, which is supported by many other researchers [Baars, Chalmers, ...etc]. Medical researchers have been able to examine specific behaviours and show how these behaviours change when the brain is stimulated or restricted (for example using Anaesthesia).

Philosophical arguments, like Chalmers and Dennet's, seem to start with the idea that we know everything about a world and then attempt to explain a phenomena. These are worthwhile mental exercises, however the problem is we don't know everything about the world so taking that kind of approach I suspect leads to circular arguments. It sounds to me that we need a better problem definition before (or at least at the same time as) we figure out how to solve the problem. And again to make sure we are explaining the right thing. The point is we won't necessarily be able to explain the function consciousness works until we can explain what we mean by consciousness. Which explains why researchers (philosophers especially) end up arguing themselves in circles.

Chalmers does come up with some open questions (p308) which can be a good start to ask when considering the operation of consciousness.

What we can get out of this is that thus far none of the models satisfactorily explain consciousness, and according to Chalmers there is still a lot of work to do. That I agree with.

1.3.4   Christof Koch

Much of Koch's work [16, 17] discusses neural consciousness correlates (NCC), which are essentially how we measure if people are actually conscious. These techniques are brains scans looking for activity in particular locations of the brain. He discusses the historical attempts at finding accurate NCC's and the difficulties in finding reliable measurements and the need for a reliable measurement.

The Feeling of Life Itself [17] discusses consciousness with respect to psychology and neuroscience. Koch develops what he calls the Integrated Information Theory of consciousness. This theory is based around five postulates:

1. Intrinsic existence,
2. Composition,
3. Information,
4. Integration
5. Exclusion

- Intrinsic existence asserts that consciousness exists for itself without an external observer.
- The Composition postulate asserts that any experience is structured. Koch goes on to discuss how much information is involved in a system and how this can be measured. This includes the interaction between elements in the system.
- The Information postulate demonstrates that a mechanism contributes to experience when its state specifies information to the extent that it demonstrates its cause and effect within the system.
- The Integration postulate demonstrates that a conscious experience requires the cause-effect structure specified be as unified or irreducible.
- The Exclusion postulate suggests that a system will lose information if it is broken into two or more parts. As such, only the "Whole" will have experience and thus be conscious.

Introduction

One application Koch considers is consciousness in computers. A CPU architecture is examined using IIT and concludes that the processor does not develop feedback loops that contain information to create a non-zero "Integrated information" or Phi (φ). This essentially results in

These principles still seem to be fairly philosophical in nature and as far as I can tell, do not amount to a workable model. However, as with other researchers we still can use the IIT concept to examine any potential model we can come up with.

1.3.5   Bernard J Baars Global workspace theory

Baars [1–4, 22] developed the global workspace idea for consciousness from detailed examinations of brain function. The "global" part is developed from the idea that sensory data becomes available to the entire brain. The workspace concepts comes from the limited number of ideas that can be kept in memory at once. Baars backs up his concept with detailed examinations of the operation of the human brain. It is shown that many of the features described in the global workspace model are present in measurements of the human brain. The basis of Baars theory is developed from the fact that we can tell the difference between waking, sleeping, comatose and vegetative states, from measurements of the brain using EEG (Electro-Encephalograms), and FMRI (Functional Magnetic Resonance Imaging).

General Description of GWT:

The basic version of the Global workspace model has 5 components:

1. The Stage of Working memory (the workspace)
2. The Bright Spotlight of Attention
3. Actors competing for the Spotlight
4. Contextual Frames and scene shapes
5. The Audience

According to Baars descriptions sensory input is delivered to the Stage as working memory. The stage represents working memory. The sensory data is highlighted by the spotlight. The spotlight is essentially a data mask that limits the amount of data that needs to be processed in working memory. The various actors examine the sensory information and compete for attention in the spotlight. The actors represent individual processors of the sensory information. The background frames collect context for the sensory input. The Audience represents long term memory.

While I am not totally convinced of the interaction between the components, however I agree that these components are essential in the operation of consciousness. I'll discuss this more in Chapter 0.

Introduction

Baars reviews the development of cognitive theories from early psychology and brain surgery, through more modern theories. It is noted that behaviouralism from 1913 and other theories, especially in psychology of the time, entirely denied conscious experience. It's possible that I am being simplistic but, much of this seems to be a rejection of previous theological explanations of consciousness and "the soul". As such the study of consciousness was discouraged in society until later eras.

Many of the references Baars examines refers to brain surgeries performed from 1920 to 1950 in attempts to treat epilepsy.

Throughout the book a very detailed account of the known structure an operation of the human brain is examined. The central structure of the brain that Baars associates with consciousness is the Extended Reticular Thalmic Activating System (ERTAS). This is an area of the brain involving the Thalamus and Reticular formation. These areas of the brain appear to be central to the operation of wakefulness, orienting response, focus of attention, or alerting the content of consciousness. These areas are the most central areas of the brain and appear to have a high degree of connection to many other (if not most) other areas of the brain. These areas of the brain primary areas that control consciousness.

With the advent of EEG it became possible to directly measure conscious states. After this point scientist and doctors could study consciousness without invasive surgery. The development of MRI and allowed high resolution imaging of the brain and has revealed a lot of detail about the structure and operation of the brain.

Baars presents a total of seven models of global workspace with increasing complexity and capabilities. The basic concept of global workspace theory defines GWS using three processes: Competing inputs, the global workspace and receiving processes. The most complex model includes the operation of frames to support goals, decision making and the "self" concept. Each of the chapters adds to the model and examines in detail the operation of the model and the supporting evidence for the model's operation.

Baars gives a set of questions that he feels are unanswered by current theories of consciousness and any new theory must address. The first group of questions relate to the book itself and Baars does a reasonable job of discussing these questions in the relevant chapters.

- Why do we lose consciousness of habitual stimuli and automatic skills?
- Why are we unconscious of local perceptual frames that help shape our conscious percepts and images?

- Why are we unconscious most of the time of the conceptual frames, (and) the presuppositions that form the framework of our thoughts about the world?
- Does the common Idea that we have "conscious control" over our actions have any psychological reality? Is there a relationship between consciousness and volition?
- What, if anything, is the difference between consciousness and attention?

The following questions are still unresolved by Baars.

- How does the item limit of short term memory fit in with a globalist conception of consciousness? We know that with mental rehearsal people can keep in mind 7 plus or minus 2 unrelated items – words, numbers, or judgement categories. But that fact does not fall out of the GW framework.
- Why do perceptual and imaginal processes have a unique relationship to consciousness? What is the difference between these "qualitative" and other "non-qualitative" mental contents?
- We are never conscious of only single features or dimensions, such as specialized processors presumably provide, but only of entire objects and events, that is, internally consistent and complete combination of dimensions. Why is that?
- When we say "I" am conscious of something, what is the nature of the "I" to which we refer? Or is it a meaningless common-sense expression?

Many of the examples and investigations of the models examines in detail the use of and interpretation of language. Much of the structure of global workspace model uses language as a basis. A description of how language is processed is given in terms of the internal dialog of several processors. These processors examine different aspects of language processing including Lexical, Syntactic, Semantic, and Pragmatic. Each processor asks different questions of a sentence to determine the meaning of the sentence. [It's also a funny sentence.] These may not be the only processes required to process language, but it does demonstrate the concepts of specialised processors acting in cooperation from a workspace. I'm not sure if this would need to be "Global", but the example is an effective demonstration.

This is certainly instructive on how we process language, but my feeling is that language is not a fundamental feature of consciousness. That said from the description and other definitely need consciousness to be able to understand language, so there is a definite relationship between language and consciousness.

Baars uses, what is referred to in Psychology, as frames as the basis for modelling structured thought, specifically goals. Frames are defines as the information that nervous system has already adapted to. This includes both conscious concepts and unconscious reflexes. In the global workspace model frames are used to define goals and how they are achieved.

Conscious rivalry occurs when two streams of sensory data provide conflicting information one stream is ignored by the brain. This can be demonstrated easily from looking at visual stimulus from both eyes. If the visual field is set up in experiments where one eye is given a completely different image than the other we find that only one image is perceived at a time. The vision can flip between the two images but both images are not display at the same time. This demonstrates that we can only be conscious of one side of conflicting stimulus or even conflicting ideas. This phenomena has been noted by many other researchers (including [14], [17], [20], [21]).

Baars notes in several chapters that repeated stimulus tends to fall out of consciousness. He details several experiments where people receive constant or repeated stimulus and measures their reactions, both physically from pressing a button and via EEG brain measurements. This might be surprising for people. But it is an important quality we need to be able to integrate into any new model.

From his work the two concepts I have trouble with are frames and the idea of self. Frames are described fairly vaguely, they are defined as a system that shapes conscious experience without itself being conscious. As I understand it frames are the background information that allows us to make connections and relationships between concepts. I don't disagree with the concepts as presented, but I think these areas can be explained much further. The idea of frames may be an inherent property of the knowledge as opposed to a structure of the brain that supports the concept. Either way the concept of frames must be supported by the operation of the brain even if there is no hardware that keep track of which frame we are using.

I also find that the description of "self" is very limited. The self-concept is based from the psychological idea of self. I found the description fairly vague. Baars refers to the work of Dennet to say that "self refers to that which I have access to". The discussion on this point seems to take the psychological view, which does not seem to be well defined in psychology. Many researchers still consider the idea of self a "delusion of common sense" (references). Baars states that there is much research on the topic, but I found the overview given somewhat lacking in explanations. This gives me the impression that the idea of self still needs a lot more work.

That said Baars work provides a detailed account for the biological operation of the brain and how this relates to consciousness. As many other researchers in the field note Baars work is considered fundamental for research into consciousness.

1.3.6 Peter Caruthers "The Centered Mind"

Caruthers describes consciousness as a sensory based processing system. He demonstrates that working memory, memory in general and even abstract (amodal) reasoning are generated from sensory based understanding.

Caruthers examines various points of consciousness including how memories operate, from short term working memory, to long term memories, Structures of Consciousness, memories, sensory information, beliefs, goals, decisions, and intentions. It is demonstrated that the majority of these have a sensory basis in theoretical models along with measurements of the brain and other experiments demonstrating how people use these capabilities.

One of his largest discussions is on alternatives for sensory based consciousness or the amodal operation of the mind and compares theories about how this would operate. Concludes that amodal consciousness, i.e. without sensory information, is unlikely.

Carruthers reviews the work of Baars on Global workspace theory and again demonstrates the evidence that the signals from our perceptions are transmitted globally around the brain. The work of many other researchers is mentioned (Tye, 1995, Block 1995, Dennet 1978, 1991, Sergent et. al. 2005, Dehane et. al. 2006, Bekinschtein et. al. 2009, Dehaene & Changeux 2011) describing theories consistent with GWT along with reviews of associated evidence from brain measurements and experiments.

Alternatives theories for the sensory based model are reviewed with the same detail as those for the supporting evidence. It is concluded that the alternative theories are not consistent with the sensory based model.

In the chapter "Evolution of reflection" Carruthers considers the similarities between human consciousness and the apparent ability of many animals to have many of the same capabilities as humans including tool usage, planning and speech, even if limited. This suggests that many animals have a similar structures for working memory and consciousness in general.

Caruthers presents well developed examples supported by research and compares his preferred models with alternatives found in research. His research presents a convincing argument for the sensory basis of the operation of the mind and consciousness, but stops short of developing a functional model of consciousness.

1.3.7 *Charles Fernyhough Voices in your Head*

{What us? ... Hello}

Charles Fernyhough in "The Voices within: The history and science of how we talk to ourselves" [22] reviews the psychology, brain function and historical references related to inner voices.

Fernyhough shows from brain scans that the speech centres of the brain include: the inferior frontal gyrus, Broca's area, Wernicke's area, Herschel's Gyrus and the superior temporal gyrus. These areas are typically activated in brain scans during when people listen, speak, or speak to themselves internally.

In these brain scans our inner dialog activates the superior temporal gyrus on both sides,, the left inferior, and Medial frontal gyrus. These areas are also related to our concept for the "theory of the mind", which allows people to guess what other people are thinking about.

He discusses children's development and how their internal speech begins development as external speech where they ask themselves questions and answer their own questions. The theories discussed suggest that as children develop these external speech becomes an internal voice that people use in much the same way, asking themselves questions and creating responses to their own questions.

When we talk to ourselves it was found that our speech network activates a region of the brain, the right temporoparietal junction, which is part of the theory of the mind system (voices within p115). This suggests that while we talk to ourselves we are using part of our brain that simulates how our mind work or potentially how other people might think. While this is useful for understanding the voices we hear in our own heads, it also demonstrates that people can simulate how we think and take a guess how other people might think.

The historical records of people hearing voices include two women (Margery Kemp and a woman known as Julian) from the medieval era who wrote journals discussing their experiences of hearing voices. These voices were often benign or helpful and the people experiencing them sought advice about whether these voices were actually the voice of god. The stories of Achilles and Odysseus are also discussed and the idea that even in mythology and ancient tales unexpected voices were often interpreted as inspiration coming from external sources of the gods. So it is expected that this phenomena has existed in humans for a few thousand years at least.

For me the interesting part of Fernyhough's work is that I kind of predicted that this functionality would exist, before I had heard of his research. This gave me a lot of confidence in writing this book. Essentially, I predicted that we would specifically have a region of the brain that produces one or more voices that is also directly connected to the speech processing part of our brains as I will discuss in Chapter 0 and Chapter 4. Fernyhough's work confirms that our brains can actually generate sounds that are interpreted by our own brains as speech. In some if his later chapters he also discusses how the brain can generate non-verbal hallucinations including visions, feelings and even smells. Given that we have the ability to generate speech it is not surprising that we have the capability to generate other sensory perceptions completely internally.

1.3.8   David Eagleman "The Brain"

{Horror movie character: "It's coming! Run you fools!"}

Eagleman [23] gives an overview of the structure of the mind and its operation, He discusses: the self, our perceptions- Neural Binding, Control, Decisions, Interactions between people, and Artificial Consciousness.

The Self how this develops in the brain from the growth and pruning of Synapses. A similar process can occur in AI Neural Networks. Unneeded links can be trimmed back. We are born with some unconditioned reflexes that are essential for survival.

It is noted that teenagers tend to take risks more often and that male teens brains tend not to be fully develop until in their 20's. Brain plasticity allows changes in the structure of the brain which allows learning and accumulating memories. How these combine to form Sentience and how this relates to the mind body problem.

One of the ideas Eagleman is most noted for is the Binding problem: How the brain synchronises sensory systems and variations in the sensation of time. This also relates to the plasticity of perceptions is discussed including the integration of artificial sensors and problems that occur with perception including Synaesthesia, Schizophrenia. It is noted that our perceptions are used to create an internal simulation that creates our expectations of events in the outside world.

Consciousness and Self-Control is considered, from examining unconscious reflexes, Conscious Flow states and the feeling of free will. However, it is demonstrated using transcranial magnetic stimulation to the primary motor cortex while having a movement primed can convince people that they desired to make the action that was induced. This suggests that our consciousness is easily influenced and often give the illusion of free will.

Introduction

There are many influences on decisions. Multiple competing networks in our brains that calculate different aspects of any situation we find ourselves in. Emotions, body state (cold, hungry etc.) motivates decisions which means people tend to make arbitrary decisions.

An idea presented is that we consider or simulate the future to predict consequences of our actions. This action is managed by the dopamine systems in the midbrain. Unfortunately this mechanism can be hijacked especially by drug addiction. These problems are created and reinforced by poor impulse control. It is also shown that these problems can be addressed by training to improve impulse control using brain scans.

Interactions between people are examined by considering that people assign intentions to abstract shapes given appropriate context. It is discussed how Autism and Asperger's reduces peoples interest and ability to understand or recognise facial expressions and emotion of other people.

The pain matrix is discussed showing how social rejection causes pain across the entire brain. My guess is that the pain matrix to interrupts any thought to force the life form to focus on the perceived injury. This line of though is considered by showing that in many situations we think of other people as objects as opposed to thinking feeling beings possibly to limit our own emotional reactions. This can lead to people becoming outgroups and allows populations to accept genocide.

Artificial Consciousness is examined starting with artificial senses, such as cochlea and retinal implants. Eagleman stats that "The brain does not care where it gets sensory data from", we will learn to integrate sensory data into our model of the world. People have learned to control robotic limbs via direct brain links.

Future research is also discussing including implications of cryogenics and the idea of becoming immortal either biologically or artificially. This discussion covers the modelling the human brain in The Human brain project and implications of preserving memories.

Finally consciousness is discussed as an emergent property of the mechanisms of our brains. While Eagleman does not develop a model of consciousness himself, he does describe the brain as creating a "simulation" of ourselves and the world around us. This idea is central to the concept developed in the next chapter.

1.3.9   Gerald Edelman

Gerald Edelman focused much of his research on the biological operation of the brain with the goal of understanding consciousness. His research included "The remembered present: a biological theory of consciousness" [21], "Bright air, brilliant fire: On the matter of the mind"

[24], "Neural Darwinism: selection and re-entrant signalling in higher brain function" [25], "Consciousness: The Remembered Present" [26], and "Biology of consciousness" [27].

The operation of the brain has determined by mapping between activation of neurons between areas of the brain. One concept learned was that areas in these maps often have neurons connecting in opposite directions between areas. This allows what is referred to as re-entrant signalling. Re-entrant signalling is discussed in detailed and has been developed from his research the Theory of Neural Group Selection (TNGS) which explained how neurons group together. The feature of re-entrant signalling is thought to have a few possible uses in the operation of the brain. We are still not entirely sure of what process actually occurs. Edelman examine the connectivity of the brain and looks at the processes that occur to build up the general process that occurs in the brain. This mapping shows the areas of the brain whose function we do know and how these components connect together.

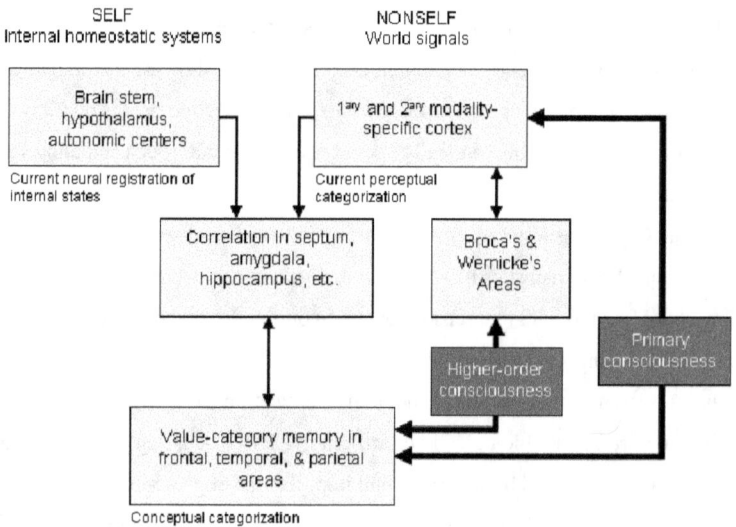

Figure 2 Edelmans Model of consciousness [24]

Perceptual Experience is discussed and how this relates to the internal operation of the brain. This relates to the Neural Darwinism schematic of "global mapping" shown in Figure 2 and from this a general operation of a human brain.

Edelman discusses language formation and the idea of "pre-syntax". We know that most animals do not use language yet they still manage to communicate. This means that we need to have support for concepts without specifically using language. The evolution of language centres is discussed along with the idea that concepts could have developed without language. It is suggested that the concepts we would most likely need are the concepts that are evolutionarily necessary.

Edelman discusses a functional model of language that is similar to that described by Baars. This structure splits up language processing into layers from syntactic to conceptual. The interaction between Broca's area and Wernicke's area is examined and compared to how his model of consciousness operates.

Edelman has developed a functional model of consciousness based on the structures of the brain. However, as Searl [10] and Chalmers mentions, it does not specifically define where consciousness emerges in the brain. That said the processes described are similar in many ways to "the design" shown in Chapter 2 of this book. As I understand it, the work of Hierarchical Predictive Coding (HPC) has developed from the ideas of Edelman discussed in this chapter.

1.3.10 *Stanislas Dehaene "Consciousness and The Brain"*

Dehaene's "Consciousness and the Brain" [28] gives a broad discussion of many aspects of consciousness, giving details of various experiments that show the nature of consciousness and the human brain. Some of the history of consciousness is reviewed along with various conscious and unconscious phenomena, consciousness concepts, binocular rivalry, change blindness, agnosia, faulty consciousness, binding and priming, out of body experiences that are commonly experienced by people.

Dehane often discusses the boundaries of conscious actions demonstrating that many actions can be performed with minimal conscious effort. From various experiments Dehaene demonstrates that the brain performs statistics, summing up the averages we find in the world and extrapolating from previous experiences. He notes that "Coincidence is triggered unconsciously, while memory tracing requires conscious effort." When people perform complex calculations in mathematics, the calculations are performed consciously, however simple mathematic operation can be performed unconsciously. Priming actions by seeding initial related stimulus (such as a similar word) allows actions to be activated unconsciously and much faster than completely unconscious or fully conscious actions. Much of the brains activity is shown to be "priming" where neuron clusters are activated, but not necessarily conscious.

Another major we consciousness is used is the Social Sharing of information. Consciousness seems to be required to understand what other people are communicating and engaging in conversation.

Introduction

The recent research is presented on the presence of the P3 wave and its implications. Various experiments are discussed showing the edge of Consciousness occurs lags events by as much as 300ms. At this point we see an "Avalanche of Consciousness", where the "P3 wave" occurs. The P3 wave shows a Cascade of activation, from sensory inputs moving towards the prefrontal cortex. In terms of information in the P3 wave appears to be an inhibitory signal that deactivates unneeded or unwanted activations in areas of the brain.

The model Dehane follows is still based on Global workspace although the idea has been expanded and validated with his and others research. Dehane's work is aimed at improving medical understanding that may produce much more accurate detection of consciousness, diagnosis of damage to consciousness and more effective clinical interventions for diseases and trauma., While this interesting in its own right, it is not the focus of the developments presented in this book. That said much of the research presented by Dehane is a fascinating examination of what we know of consciousness.

1.3.11   Hierarchical Predictive Coding

Hierarchical predictive coding has been developed as part of cognitive science in an attempt to explain perception. The general process requires a model of the world or an object to be matched with incoming sensory data. The errors in the sensory input can then be used to learn new information about the previous models.

The HPC process combines two processes: bottom up and top down processing. The Bottom up process analyses sensory input and decomposes the sensory data into classifiable structures. The top down process uses the classified structures to rebuild the incoming sensory data. Between these processes an error minimisation process occurs so that the classifiers used improve their representation of the incoming sensor data.

Introduction

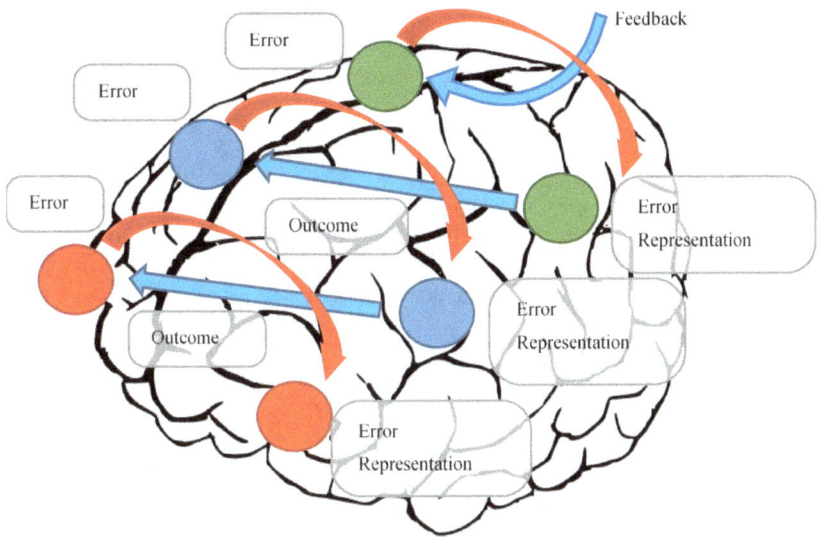

Figure 3 Hierarchical Predictive Coding Bottom-up Processing

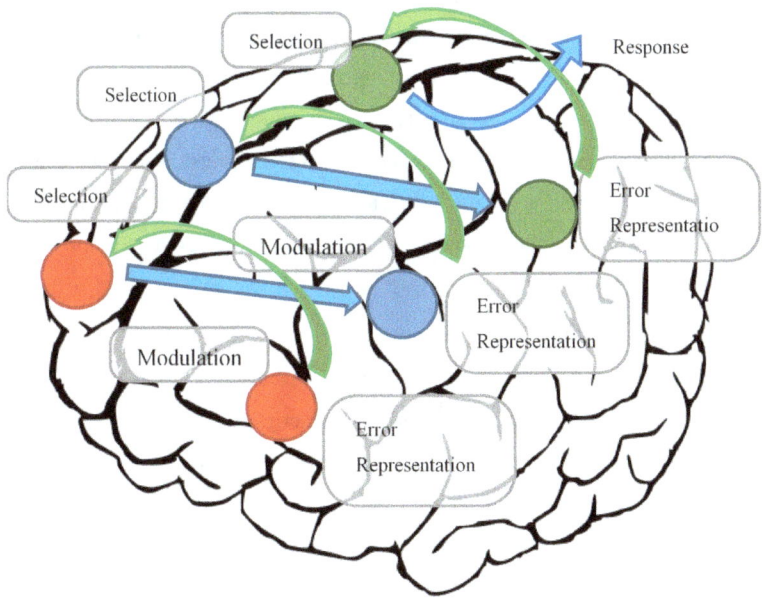

Figure 4 Hierarchical Predictive Coding Top-down Processing

Introduction

HPC has been successfully applied in medicine analysis of various brain conditions and operations ([28]–[33] and many others), and in more computational fields to develop or explain various processing phenomena ([34]–[37] and many others).

Many of the computational systems have used 2D image analysis successfully. One of the limitations in the research that I can see is that humans have two eyes which results in stereo vision and depth perception. My first job out of university was with a company that developed photogrammetry software that used stereo vision to generate 3D models. 3D Models can be used to examine the relationship between objects in 3d dimensions, specifically this can be used to determine occluded objects.

I suspect that the integration of stereo vision into a HPC model may help explain some of the difficulties identified in HPC research. The design I propose here is similar in many respects to HPC in operation, except that I have used 3d modelling instead of 2d.

1.3.12   Bayesian Markov Models

A Bayesian model is a statistical model where probability is used to represent all uncertainty within the model. The Bayesian Theory of Mind (BToM) attempts to determine beliefs and desires (internal states or an internal model) based upon their output actions. The Bayesian Theory of Mind (ToM) framework has become a common approach to model reasoning about other agents' desires and beliefs based on their actions [38]–[44].

These models are often used to better simulate human-like behaviour in artificial intelligence. These methods have been applied to such tasks as path planning and exploration. Potentially BTOM is a method for attempting to understand another person's beliefs or desires, which means we might employ similar reasoning structures to accomplish the same task.

1.3.13   Psychology

The previous research presented on consciousness seems to be used as the idea of consciousness in psychology. As such much of the descriptions from Psychology use the work of Baars, Koch and others. From what I have seen psychology seems to focus on the content or operation of the mind, along the lines of "this is what people think". Whereas what I am interested here is the structure and functions of the mind that contribute to consciousness, where do the signals in the brain occur, what data is input and how the data is changed to contribute to consciousness. The previous sections have primarily covered the structure and function of the brain from a biological and often a philosophical perspective.

So the main structural models of the mind that I can find was the famous psychoanalyst Sigmund Freud. Freud developed a model for the structure of personality including the Id,

Super Ego and Ego [45]–[49]. The idea is that these three components of our personality are in constant conflict which ends up leading to many of the psychological issues people experience.

The id is described as the most primitive of the three structures. It is concerned with instant gratification of basic physical needs and urges and operates entirely unconsciously. This would include basic biological behaviours such as being hungry when we see food or, Freud's favourite, being sexually attracted to someone.

According to Freud the superego is concerned with social rules and morals, similar to what we think of as our "conscience" or "moral compass." It develops as a child learns what their culture considers right and wrong.

An example of how the Superego and Id interact would be to consider someone walking by with an ice-cream. The Id would desire the ice-cream and potentially snatch it by the superego would suppress the urge as it would be considered rude. However, if your id was strong enough to override your superego's concern, you would still take the ice cream, but afterward you would most likely feel guilt and shame over your actions.

The ego is considered the rational, pragmatic part of our personality. According to Freud, it is less primitive than the id and is partly conscious and partly unconscious. It's what Freud considered to be the "self," and its job is to balance the demands of the id and superego in the practical context of reality. So, if you walked past the stranger with ice cream, your ego would mediate the conflict between your id ("I want that ice cream right now") and superego ("It's wrong to take someone else's ice cream") and decide to go buy your own ice cream. While this may mean you have to wait 10 more minutes, which would frustrate your id, your ego decides to make that sacrifice as part of the compromise– satisfying your desire for ice cream while also avoiding an unpleasant social situation and potential feelings of shame.

Freud believed that the id, ego, and superego are in constant conflict and that adult personality and behaviour are rooted in the results of these internal struggles throughout childhood. He believed that a person who has a strong ego has a healthy personality and that imbalances in this system can lead to neurosis (what we now think of as anxiety and depression) and unhealthy behaviours.

There was and still is still some disagreement about Freud's method, however his model of the mind still seems to be at least accepted in the field of psychology.

The idea of self is related to consciousness, for example self-consciousness. While there is extensive literature on the topic ranging from self-image or identity, self-esteem and self-ideals. There are literally thousands of research papers and books on the self. Lewis [50]

"Self-knowledge and social development in early life", Baumeister made this a focus of his research in "The Self"[51]. Both of these books discuss many aspects of the self from a psychological point of view.

Sadly, there does not seem to be a coherent model of consciousness that has been agreed upon in psychology. That said the purpose of psychology is to understand the operation of the mind, not necessarily the underlying functional structure. As such I can understand why such a model has not emerged. *{And he still needs therapy so he's trying not to piss off the entire psychology community.}*

1.3.14 Summary of research on Biological Consciousness

Despite the fact that most of the theories of consciousness were developed nearly 30 years ago there has been very little progress or alternative theories developed to explain what consciousness actually is.

There is a lot of research around on consciousness, according to google scholar over 600,000 research papers. This is a lot of information for any one person to spend a lifetime consuming. Given that I have not read every paper that exists I have probably missed quite a bit of it. Hopefully I have covered the majority of the main theories people work with currently. Apologies to anyone whose work I missed … there are a lot of you.

That said we have established where much of the biological processes for consciousness are in the human brain. And there is a lot of work showing the details of where specific thought processes occur and what happens to people when they have injuries or some other form of brain damage.

The limitation of measuring brain activity is that we don't actually know what is being processed. We can't at present decode brain activity to examine exactly which sensory data is being examined and what kind of processing occurs.

So the alternate strategy is to look at artificial consciousness. The advantage of artificial processes is that we can determine exactly what data is being examined and how it is being processed. That said we still have not found any examples of artificial consciousness.

1.4 Review of AI Research

Current work on consciousness has had some success describing how the brain works, how this produces a mind, and some specifics of the information processing that occurs. As yet a model of consciousness has not been completely agreed upon although, Global workspace theory GWT and Hierarchical Predictive coding HPC are the front runners as far as I can tell.

Introduction

One of the biggest gaps I have noticed in the research is a full systems model of these processes. This is one approach I have some experience with, as I will demonstrate in Chapter 2. The following chapter contains fairly basic descriptions of a broad selection of AI techniques. There is far more depth that can be studied on these topics each of which could and does fill several other books. If you do find these topics interesting I would recommend reading on these concepts specifically. That said I am attempting to give an overview of the techniques so that we can compare them to how biology works and potentially use similar techniques in our design. This also establishes that the techniques can perform tasks similar to those found in biology.

This book is about a design I have come up with that may represent what we 'could' call consciousness. Whether it actually is or not is difficult to prove as I discuss in Chapter 7.3. This design has been developed from some research about automation in manufacturing environments. I considered that the process I had developed could be used to distinguish between a robotic arm and other objects in its environment. The idea was that a three dimensional simulation was used to describe an environment and synchronised with real-time scans of a real robotic system. This concept was able to identify differences between the simulation and the real environment. I considered that the simulation of the robotic arm when compared to the scan of the environment could be considered self-awareness. This would mean that self-awareness has something to do with the combination of the simulation and the external sensors.

The idea of self-awareness could be expanded to general consciousness if more than a physical model can be considered. When considered further I realised it could also be applicable to people. Essentially people and other animals have different sensors, but there is no reason that these sensory inputs could not be simulated and compared to their real inputs. Could this be an idea of how self-awareness and ultimately consciousness are developed?

1.4.1   Current AI Tools

To expand on the ideas of what else could be used to explore consciousness the following section is attempting to discuss relevant techniques that are used in AI and how they can be applied to consciousness.

When I first looked at Artificial intelligence around 1997 I was really disappointed. My first AI unit and I was all "Wow, cool let's get the AI going" at the end I was really disappointed. In this AI course I learned that AI does 2 things: Searching and sorting.

Introduction

Sorting is fairly simple. Either find a particular value typically either the minimum or maximum of a value or create an ordered list of values. Sorting algorithms include bubble sort, heapsort, and quicksort among many others.

Searching is a little more complicate, but basically it requires creating a "network" of options. Where the network is considered a set of "nodes" connected through links. The search algorithm attempts to find a route through the network to a goal state by searching through the nodes and links.

Searching deals with networks or branching options and finding an optimum path to a goal state at the other side of the network or decision tree. This network we can call the problem space.

Sorting is simply getting a bunch of options and arranging them so that they are ranked in the order of some property or calculation… ultimately a number. These calculations are often called fitness functions and are used on each option in the network. All the searching does is find the best option for the next node to search in the network.

For example, in Chess the network is constructed by considering the exact positions of each piece on the board. A single move changes the state of the board. In Chess there are 16 pieces on each side. So if each piece has one move there are 16 possible choices per state. (There are typically more when all of the pieces on the board). Each successive move reveals another 16 moves. The general equation is n^{16}, so after n moves there are n^{16} possible final states. For example after 5 moves there would be 152587890625 final states. A lot of possible options for a human to examine. What AI attempts to do is search the options to find a "goal state" and attempt to search the states with a minimum effort which is where fancy fitness functions come in. The fitness function calculates is the next best state to examine to find the goal state (in Chess a checkmate) the fastest.

There are many forms of AI, but they are all used to search a network, sort options or both.

A few of these AI techniques are as follows:

- Neural networks (mainly sorting, but can get complicated enough to do searching at the same time)
- Fuzzy logic (sorting)
- Expert systems (sorting and searching)
- Knowledge representation (searching)
- Genetic algorithms (sorting and searching)
- Creativity (searching. This deals with expanding the network)

- Intelligent agents (sorting and searching)
- Perception (a different type of searching)

These techniques are all methods of sorting and or searching in some combination. They each attempt to find efficient methods for the best option in a set that is the best option in a network of options.

Learning algorithms can be described as methods to expand the network in a known problem space.

Creativity can be described as methods to expand the network in an unknown problem space.

Genetic algorithms are a method of changing a set of genes that represent part of an algorithm.

The genes represent the states or options in the network and the "fitness function" is the method of sorting the different states. These work by randomly or progressively changing the genes to produce a new "entity". Then the new entities are evaluated and a few of the "best" results are kept to produce another set of entities. This results in better solution evolving.

Intelligent agents and cooperative behaviours are generated from creating rules for a simple entity so that when combines the distributed behaviour can solve problems. Each agent can search problem space independently, in effect a parallel search system.

Perception is a bit more interesting as it looks at input data and attempts to find objects or recognisable structures in the input data.

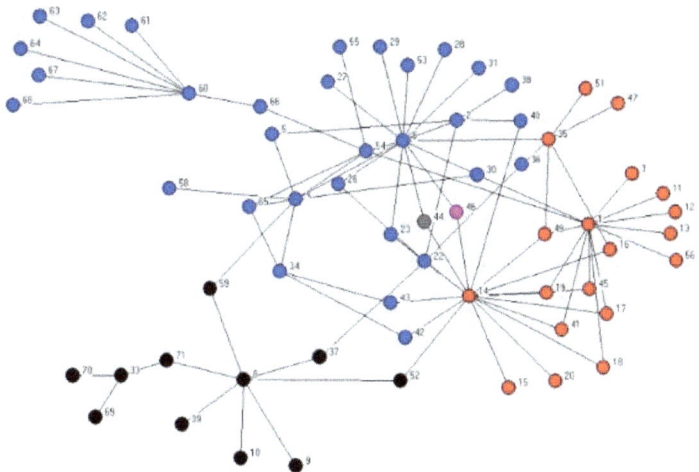

Figure 5 Network Diagram

Search algorithms include breadth first, depth first, A*, C*, D*. Search algorithms map the network then from a starting point, evaluate all of the outgoing nodes and find the "best" next node that moves closer to a desired goal. The two parts of any search algorithm is the network connection and an evaluation function. The goal is to minimise the number of nodes searched as the evaluation has a computation cost and at the same time find a minimum path to the goal state.

Examples of AIs that use network searches to determine actions include Deep Blue, Deep Fritz and Alpha Go. These are basically doing searching algorithms. The node in the network are represented by the combination of pieces on the board. Each move represents a link between nodes. In many ways all AI systems are performing their analysis using search algorithms.

1.4.2   Databases

A database can be described as a structured collection of data. Currently databases are used extensively by governments, businesses, and individuals especially on the internet. These datasets are often used to manage customers, products, shipping and other types of data.

Searching is one of the functions that are interested in terms of consciousness. The data in a database can be used to represent different pieces of knowledge. When these pieces of knowledge are connected the problem becomes a network searching system.

Databases are optimised for searching through what amounts to often millions or billions of entries. The structures of databases allows relationships to be defined between datasets. As such it is possible to look for relationships between entries. These structures are used for storage and as such are reactive. This means that they only respond when "asked a question". For example "Which people live in this city?"

This can be used to organise how knowledge would operate in a conscious entity. That said we can be reasonably certain that databases themselves are not conscious, as we have never seen a customer/product list attempt to take over the internet.

We know that people can access many different memories or pieces of knowledge in random sequences. This means that humans must perform a similar task when attempting to access their memories. So if we were to design a consciousness it would be expected that a database system would be a useful feature of designs to store knowledge and find relationships. As such when discussed in this book databases may be used as analogies for memory.

Introduction

1.4.3 Decision trees

A decision tree is a structure that breaks down conditions to allow a decision to be made. The tree structure is developed from the branching nature of initial conditions and outcomes. There may be many conditions and combinations of conditions that will produce multiple outcomes. From each outcome another level of conditions can be examined to eventually produce a structure that determines all possible decisions that can be made on a particular topic.

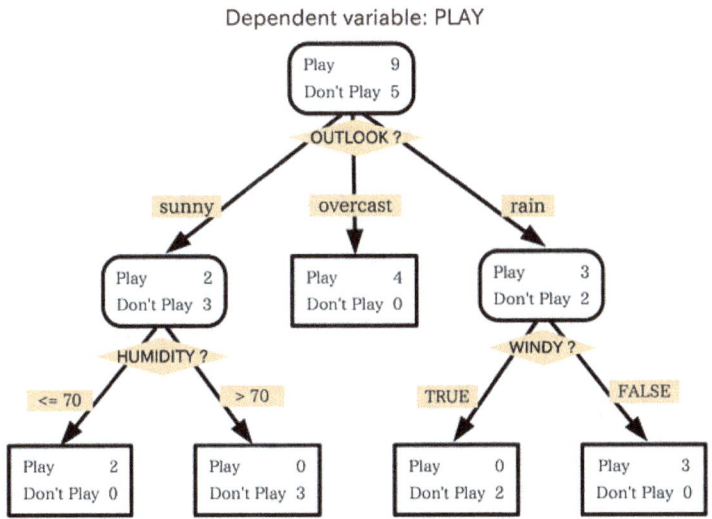

Figure 6 An example of a decision tree.

Decisions tree structures have been utilised in Artificial Intelligence to produce expert systems and are often used in games to generate behaviour trees. Expert systems generate a diagnosis from input measurements. Expert systems have been successfully generated for medical diagnosis. Behaviour trees are extensively used in computer games to simulate the behaviour of opponents to make them more realistic or simply more difficult to interact with.

The difficulty with Decision trees in that they have typically been generated manually and are often difficult to update. As such, other AI techniques have been adapted to perform the same task. That said the functionality is present in existing AI.

In terms of consciousness we know that people make decisions, however bad we might think the decisions are. As such again the same functionality is expected to be available in biological beings.

1.4.4 *Fuzzy logic*

Fuzzy logic has been developed from the idea that sometimes logic is not always binary. As such, the output/s of a Fuzzy logic system are analogue values as opposed to binary (true /false).

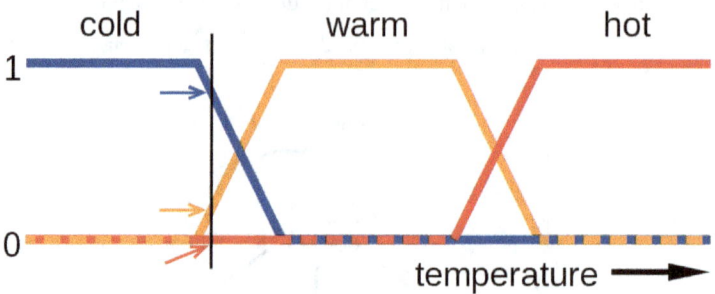

Figure 7 An example of a Fuzzy logic system

From Figure 7 we can see that temperature can be classified as "Cold", "Warm" or "Hot". The difficulty is that these outputs overlap. So for a single input value multiple outputs could be present. This technique has been successfully applied to complex robot controllers and other high speed low level processes. The advantage with fuzzy logic is that the "Fuzziness" of the controller can be adapted which allows a fuzzy system to learn an appropriate strategy for the system it is controlling. Fuzzy logic has not been widely utilised in industry, but the research concept has been adapted for use with other techniques, especially neural networks or deep learning.

In terms of consciousness, people are often faced with decisions that do not have discrete outcomes. For example which direction to move when moving away from a threat. The direction itself often has a useful range of possibilities. Fuzzy logic is a technique that could be applied in these cases.

1.4.5 *Intelligent agents*

An intelligent agent system or a distributed intelligent agent system is a collection of individual agents that are tasked to solve a particular problem. The idea with intelligent agents is that collective intelligence can be more effective at solving problems than an individual intelligence.

We can think of this in terms of network searching as many agents being created to search through the network to find a path through to a goal.

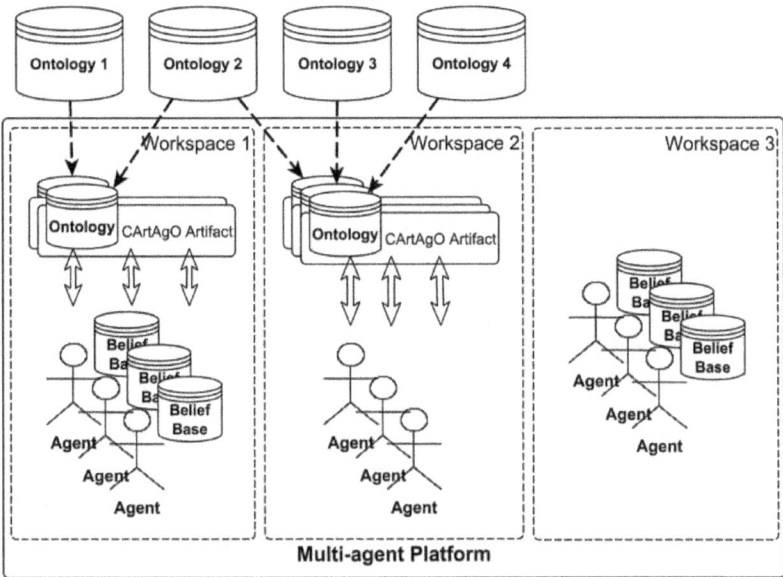

Figure 8 An example of a multi-agent system

One example of an intelligent agent system is a web search such as google. The idea is that every web page exists on separate computers that are distributed across the world. If each computer is tasked with searching its own data for specific terms of a web search they can all collectively produce appropriate results,

We can think of human beings as individual intelligent agents operating collectively in the world. We can also think of the neurons in our brains as agents performing a similar operation. Notice here that Dennet's work on consciousness comes from distributed intelligence. Also Baars idea in global workspace that signals are distributed globally across our brains, is very similar in operation to a web search. While I don't think that the wiring between neurons allows complete connectivity a search request (similar to a database or web search) sends the request across the brain so that any relevant connection can return a result to answer the query. Again much of the operation of intelligent agents might relate to how our brains work in general and specifically how memories work in biological lifeforms.

1.4.6 Genetic algorithms

Speaking of life forms we get to genetic algorithms, which obviously have a biological inspiration. Genetics is the basis for evolution. So if you are interested in how evolution works take a look at genetic algorithms. Every successive generation of combinations of genes produces a different output. The genetic algorithms use 'genes' as the data network and search though the network by implementing a complete gene set and applying a fitness function. A fitness function calculates how well a particular gene set is at accomplishing its task. The task could be purely computational or it may involve real-world processes. Either way the fitness function determines how well the gene set accomplishes the task and is used to generate the next generation. Often this involves simply randomising the genes and testing the new generation.

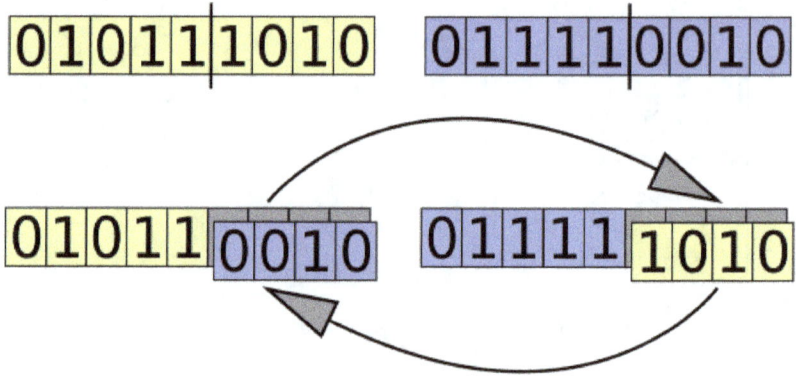

Figure 9 Simple propagation from genetic algorithms

Genetic algorithms are useful at exploring "information spaces" that have an extremely large number of inputs. Each gene operates as an input to the system and with some adaption occurring with each generation over time a solution may be found. The main limitation of genetic algorithm is determining an effective fitness function.

1.4.7   Object Recognition

Object recognition is one of the most difficult problems in AI research. The goal is to be able to recognise particular objects from input sensor data, typically images, but more recently 3d data. The basic strategy is to decompose images or 3d data into small "features", then identify specific objects from a collection of features. Object recognition can be describes as the matching process that decides which features to map to an object template. As with many tasks attempted in computer science, Object recognition is a task that people can perform fairly well, but computer science has not yet found a completely satisfactory generalised solution.

We know that people can recognise objects in the real world and perform some of the same operations to determine some times of objects. In humans we specifically recognise faces, presumably there is a similar technique where we look for the combination of eyes nose mouth and other facial features to recognise both a general face and identify specific individuals. Facial recognition is one task that has been fairly successful in computer science. This technique has been applied to facial recognition in security cameras or in a much more structured environment with passport or security ID validation recently added to smart phones.

Notably there are some slight differences between how people work and how computers recognise people. As such, it is still possible (although difficult to accomplish) to fool both a human and a machine with a carefully constructed reference image (passport or ID). This shows that there are some differences between how people and computers recognise individuals especially, but we know that both computers and people can perform the same task.

I suspect that part of the operation of consciousness is to be able to recognise and reproduce stimulus from the knowledge base of our memories, but this will be discussed further in the next chapter.

1.4.8   *Neural networks – Deep Learning*

Neural Networks (NN's) are biologically inspired from how our own brain s work, so they operate in a similar manner to actual human brains. A neural network is based upon the idea of using interconnected neurons. These Neurons are connected in layers from one layer of input to a layer of outputs, similar to that shown in Figure 10 Neural Network. The link values or weights between each neuron in the network is developed so that when a set of inputs were transmitted the network could signal through the hidden layer to generate an output signal.

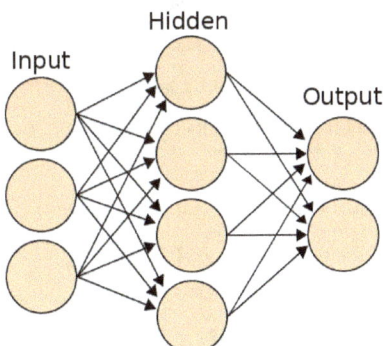

Figure 10 Neural Network

The Neural networks needed to be trained using a technique called back propagation and sets of data called training sets. The training sets are used as the inputs and the output is defined for the training set, this allows the neuron links to be calculated for the training set. Then when new images are tested the "best output" is generated from the network.

Neural networks were developed from the 1990's to the 2000's and stagnated after this point. The problem was with multiple hidden layers, neural networks would fail to generate effective links between input and output layers. Enter "Deep neural networks", which solved the problem where the link values were not transmitted between two to three layers. One of the major leaps for neural networks was the development of computer processors from NVidia that allowed NN processing to occur on graphics cards and significantly improved performance.

As Neural networks have a similar structure to networks and decision trees they can be modified to perform the same tasks relatively easily. Using convolution networks can also produce outputs similar to these described by Fuzzy logic. The advantage of this is that the network is trained rather than hard "coded". They are popular in computing as their flexibility allows them to perform many different types of applications. This may also explain why we as humans have developed similar structures to perform our own calculations.

We know that biological neural networks can produce a conscious being. While this may provide much of the motivation into neural networks in computer science, to date no-one is claiming to have produced a conscious neural network. This suggests that whatever we are doing with NN's, consciousness requires something different.

To be able to understand how the brain works we may need a better understanding of how to decode neural networks and also how to measure the connections and weights between neurons in the brain. As one could imagine measuring the weights in an actual individual neuron from a biological brain would be quite difficult. The individual neurons need to have their weights measured and the relevant connections between neurons would need to be mapped.

This process can be performed on artificial neural networks and looks something like Figure 11. This shows a sample NN and associated output that analyses the connection weights between neurons. The Blue and yellow dotted lines show which connections between neurons which represent small feature image patches and the connections between layers. These build up over the hidden layers to produce recognisable features of the output.

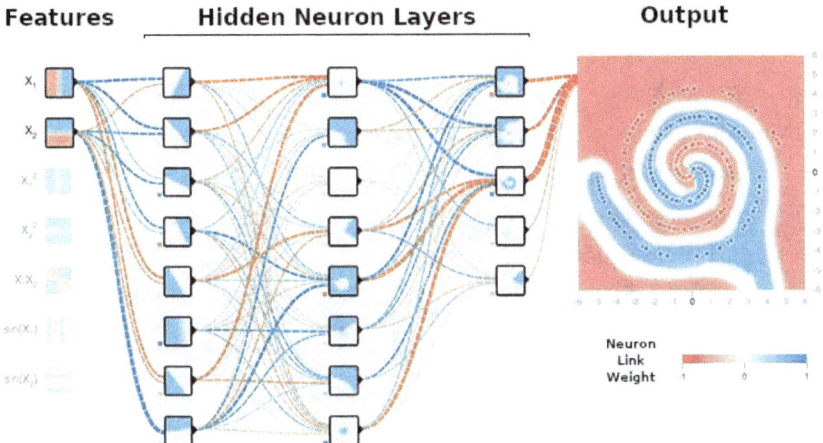

Figure 11 Neural Network neuron link weights

I suspect it will be a long time before we are able to do something similar with the human brain given the complexity. However, this process would allow us to determine exactly which connections are relevant for any process in the brain. That said we can assume that a similar process is occurring in the brain.

1.4.9   Artificial Learning

Some AI techniques have inbuilt learning functions in them. Neural networks, Fuzzy Logic, and Genetic Algorithms are examples of these techniques. These techniques learn by adjusting connections between existing nodes or internal values. These are effective at performing analysis on specific types of data, but often have difficulty with datasets even slightly outside their training realms. Basic text recognition and speech recognition are two applications that have been found to be suitable for these types of learning systems as they can learn new words "discrete entries" by adding them to a database. These are fine for basic tasks that can be represented using a few continuous (1.0 – 2.0 and so on) or discrete (1, 2, 3 and so on) variables. However knowledge can be much more complicated which requires more sophisticated tools to manage.

What we have used in learning systems such as Fuzzy logic, genetic algorithms, or NN's is that they require a "variable field" to determine appropriate connections between inputs and outputs. A learning system in this case adjusts part of the field to create new connections between inputs and outputs.

Other systems learn by accumulating data and determining connections between existing and new datasets. This requires a more abstract learning process as shown in Figure 12. What we see from examining Figure 12 is that the learning system is attempting to identify "underlying rules" from external information. What we know from our own brains is that we often require combinations of techniques to be able to learn different types of information or different types of skills. So a more general learning system would require a combination of database discrete searching along with more fuzzy logic and probably a range in between for different types of information. The underlying rules would need a conceptual structure to be able to map rules and generate data for recognition.

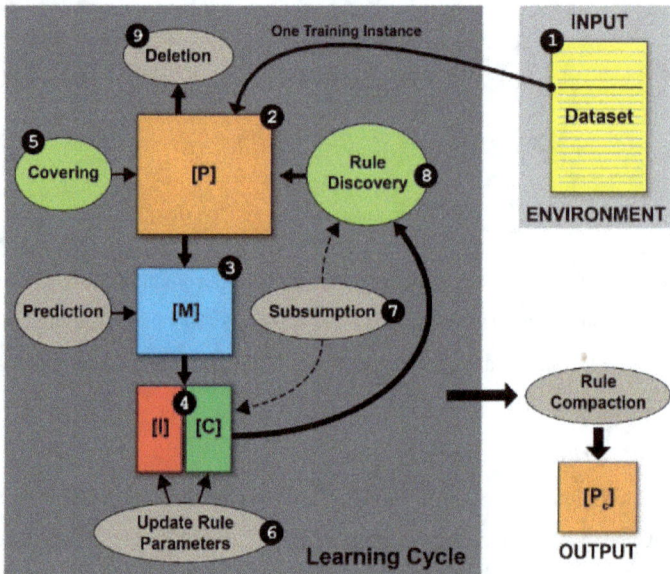

Figure 12 An example of a learning architecture

1.4.10   Generative Adversarial Networks (GAN's)

Generative adversarial networks were developed by Ian Goodfellow in 2014 [52]–[55]. GAN's take the deep learning system a step further where in many cases the NN is used to play games against itself. This allows the network to be trained on a much larger variety of training data. GAN's have been successfully applied to learning how to play games including "Chess"and "Go" [56]–[58] to older arcade games like "Snake" [59] and even "Doom"[60], [61].

This process allows a neural network to generate its own training sets by playing against itself many more times than it could against a human being. What we might find is that people actually do the same thing. Fernyhough noted that in children we develop our internal voice by asking questions out loud and answering them. This suggests that as we grow up these voices are internalised and we talk to ourselves often from the point of view of an opponent or a critic. This allows us to consider an opposing point of view and adapt any strategy we have taking into account the perceived criticism. This is in effect an adversarial learning system.

1.4.11   Consciousness in AI

There are a lot of tools that have been developed for AI. Many of these tools can be used in combination. These tools have been proven successful in many applications, from text and voice recognition and more recently image synthesis.

One big accomplishment that the current AI tools have not managed is to become conscious. Many people feel this is a good limitation, as they expect that a conscious AI would become more of a problem for people.

My feeling is that people are conscious and cause themselves more problems for themselves than AI could. So I am less concerned about AI applications than I am about psychotic idiots in the real world. That said AI's seeking world domination is a worthwhile concern. {*Evil Scientist: and fun to think about*} As such this topic is discussed further in Chapter 7.

1.5   *The closest we have got to Consciousness*

The AI's mentioned and the techniques used are most likely not conscious, they are simply performing calculations that lead to a result similar to or better than what a human could do.

Watson was an AI developed for playing jeopardy. The Ai used fitness functions to determine which pieces of information were relevant to the questions being asked and a few simple rules to create the answers in the form of a question.

The technology to develop Watson has been implemented in software tools we find in our smartphones and PC's these days in assistant programs such as Siri and Alexa. These again use speech recognition, internet searches and speech synthesis to interact with people. Often these can produce outputs such as booking an appointment where a person at the other end of a phone conversation is not aware that the call came from a computer program. In this way we could argue that these programs can pass a Turing test. These systems have the potential to construct a knowledge base about themselves, but I don't think they are conscious at least not conscious in the way we would think about another person.

The most popular AI strategy at present is Deep Learning, which is simply neural networks with more than 3 layers of neurons. One of the biggest limitations of Deep learning in particular is that they do not develop systematic understanding of their applications. What they end up with are heuristics based on known datasets. Even with their limitations these software systems are capable of performing analysis n complex datasets far more rapidly and accurately than a human being could. That said these programs are again not conscious. *{Thankfully}* These types of systems are simply determining a relationship between datasets via heavy mathematics.

There have been a few attempts to create humanoid robots, but most have not attempted to investigate consciousness. CRONOS was an anthropomorphic mobile robot designed by Owen Holland among others [62]. It used 3d printed bones and compliant joints which reduced the potential for damaging itself or humans it interacted with. It featured humanoid arms with a complex spine, but used wheels for mobility and a single camera. Holland also investigated the idea of virtual consciousness [63] and how this could be achieved using the simulation and real world sensing. That said by his own admission his constructs do not appear to be conscious.

I suspect that we will not have general intelligence until AI's can develop systematic models from datasets. A systematic understanding or model will allow behaviours to be predicted that may not be found directly from training data. As such we might find that a systematic AI will be able to out-perform a deep learning heuristic at least in terms of accuracy, provided that the model developed is accurate. Even then we might need a completely different approach to develop a conscious computer based system.

1.6   Design Goals

The goal of this book is to develop a design that could be considered conscious. From the research presented in we can extract a few other features that are most likely necessary for a design of consciousness to operate. To this end we need to consider what consciousness is and what it isn't. As mentioned in section 1.2.9 despite the breadth of research over the years the topic still eludes a consistent description even in psychology where consciousness is a fairly central concept. So what we will do in the next section is discuss the ideas and see if we can pin down a good definition of what we are attempting to achieve.

In his lectures John Searle talks about Ontological and Epistemic knowledge as part of his discussions on consciousness. Ontological has to do with existence, it describes objective facts that exist without a subject to experience them. Epistemic is to do with knowledge. This implies that knowledge is subjective and often refers more to opinions and internal experiences. They are subjective and must be "experienced" by the subject.

Introduction

So given these two concepts we can see that a conscious mind needs to be able to experience things internally, i.e. have epistemic experiences and it needs to be able to understand that external facts exist about the world. We can be sure of our own experiences at least to the point that if we are conscious we can be sure that we are having experiences, not that the experiences are necessarily accurate. Our experiences are epistemic, basically by definition.

Ontological "facts" exist externally from our consciousness, but the things we know about the real world outside ourselves must be represented in our own consciousness. Facts may about a range of concepts from the existence of real world objects all the way to abstract concepts. As such both abstract and concrete knowledge must be supported by the consciousness. Abstract thinking can be anything from numbers, language, theories anything that does not have a real world instantiation. This allows people (or any conscious being) to think about concepts that do not necessarily have a physical existence. In terms of abstract and concrete knowledge we can think about the difference between counting numbers and counting money. Numbers are abstract, there is no "one" in the real world. Money is has a real world physical representation i.e. you can have a $1 coin. That said both of these concepts must be supported in consciousness.

We experience consciousness as a singular level of awareness. However, we notice several different levels of Consciousness: including conscious, pre conscious (formerly sub-conscious) and unconscious. These three levels of consciousness relate in many ways to memory. It is clear there is a wide agreement among researchers especially Barrs, Koch, and Karuthers that memory is seen as a central feature of consciousness. Baars especially, has shown that short term or working memory is a central feature of the global workspace model. It is also clear that working memory has significant limitations. It is commonly shown that humans can recall seven plus or minus two (7 +/- 2) unrelated items from short lists. We also seem to have much longer term memories that can recall facts or images from our childhoods a lifetime ago. Barrs especially demonstrates that working memory is relevant to the bound of consciousness. In some ways we can reform this statement into the opposite which is: "Long term memory is unconscious". Then there is "Pre-conscious" which is somewhere in between, this might be related to the relationship between working and long term memory where we are attempting to "load" a concept into memory. There are aspects of our own minds that are not directly in our working memory, but will influence our state of mind such as emotions,

The things we notice are immediate. From this we know that time is a relevant topic. Ideas and concepts can be recalled from long term memory. Our senses provide information in real

Introduction

time, but often much of this information is not relevant. This gives the basis for Baars idea of the spotlight

These topics demonstrate many of the features that should or must be present in a conscious mind: these include:

- Sensory inputs
- Memory: Long term storage, working memory
- Sensor/Working memory mask – filtering: The spotlight of attention
- Representations of real world objects and abstract concepts
- Intelligence - the ability to make calculations or projections
- The ability to learn new information

So from our AI research we can see that many of these components are certainly possible. Sensors are widely available, from cameras to heat or touch sensors etc.

Long term memory in computers can be represented with "hard drives". From a software perspective this can be implemented with a database system of concepts. Working memory can be run in "RAM". Again, software structures may be required to handle the details of how this is implemented, but this should be achievable. The items stored in the database or in short term memory would representation of the different types of concepts abstract or concrete.

Intelligence is arguably the main aspect of consciousness that artificial Intelligence has addressed. Many approaches have been discussed in the previous chapter that could be utilised to generate intelligent strategies to approaches. The general idea would be to use the concept as a general structure to store and reference data. These would often be active where for example an image of an object could be matched to a template concept or how mathematics operations work where two numbers can be combined to determine a third.

Learning is common topic of research in psychology and biology, but due to the complexity of the brain it is difficult to get an exact model of how humans (and animals) learn. The good news is that learning systems is again a topic that has been successfully applied in AI. Even with its limitations, it should be possible to use AI learning techniques should be able to be used to allow a system to learn new skills or facts about the world.

The combination of these tools it should be possible to develop a functional model of consciousness, working from sensory inputs to be able to determine relationships with knowledge stored long term in databases and loaded into short term memory where relevant. These concepts can be examined while in short term memory and in some way influence our actions.

Introduction

One of the goals of this book is to determine a set of experiments or tests to see if the design is conscious or not. Many researchers have come to the conclusion that either we are not conscious or that consciousness cannot be implemented in a computer system. However, if we can develop a model of consciousness it may still be able to explain how our own brains work. Then there is the idea from Alan Turing with "Computing Machinery And Intelligence" [64] and the most famous phrase "The Imitation Game", to paraphrase if something can imitate a process in every way it might as well be the same thing. To state it explicitly: if we can develop a model of consciousness that produces the same outputs as a conscious mind we have effectively created a conscious mind.

There are a few question that have been asked in this regard:
- Can a computer program (for example and expert system) perform a test such that it can mimic any feature of consciousness, but not be conscious?
- Can the Consciousness have experience?
- Does the consciousness have intrinsic information?
- Can the model determine if consciousness has any non-zero properties?
- Does the design support Kock's 5 postulates: Intrinsic existence, Composition, Information, Integration and Exclusion?
- Can the design produce internal sensory qualia?

We can also examine how this idea of consciousness compares to other examples of consciousness that we are aware of. We assume that we are conscious and much recent research has investigated the intelligence and potential consciousness in animals. Animals we typically think are not necessarily as intelligent as humans but they can often show examples of what we think of as consciousness. Whereas computers may be considered highly intelligent as they can solve complicated mathematical problems, but are not conscious. *{That we know of!}*

What this means is that intelligence is not necessary for consciousness. However we know that at least one type of animal can evolve to be conscious and it is not a huge stretch of the imagination to assume that other animals may have similar capabilities in terms of consciousness.

In the next few chapters we will use the previous discussions to develop a model of consciousness that could be implemented in a computer system. We will then break this system down and compare its operation with the previous research on consciousness.

Introduction

2 The Design of Consciousness

This is the important bit

The goal is to make this design consistent with the phenomena we see of consciousness in the real world. It must be able to generate the same capabilities and problems identified by previous researchers. Also this design should not conflict with observations of other researchers.

The starting point I am using is the idea that to be conscious we need to be able to "simulate" our experiences internally and match the simulation to what we see in the real world. For this to be able to operate other components need to be in place and must also produce phenomena that we see in the mind.

The main notable difference between this and other attempts to define consciousness is that this attempt is aimed at a processing based explanation that could be implemented in an artificial system, but it should also be consistent with the biological phenomena we see.

Confused?

Well, it is a lot to take in at once especially if you aren't used to thinking about system analysis and design. So this next section breaks down this diagram and explains both the individual components and the main processes involved.

The Design of Consciousness

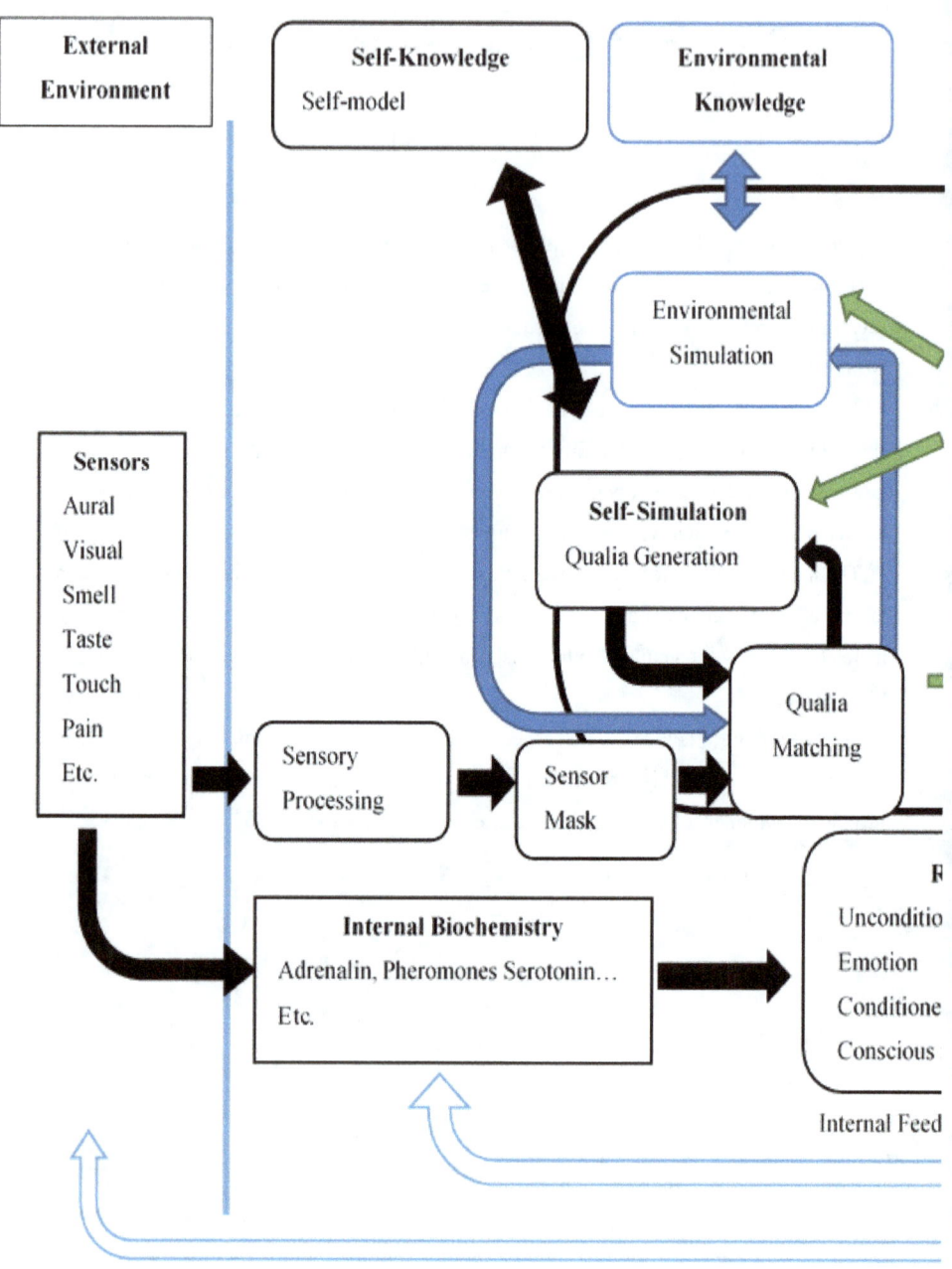

Figure 13 Design of Consciousness expanded Part 1

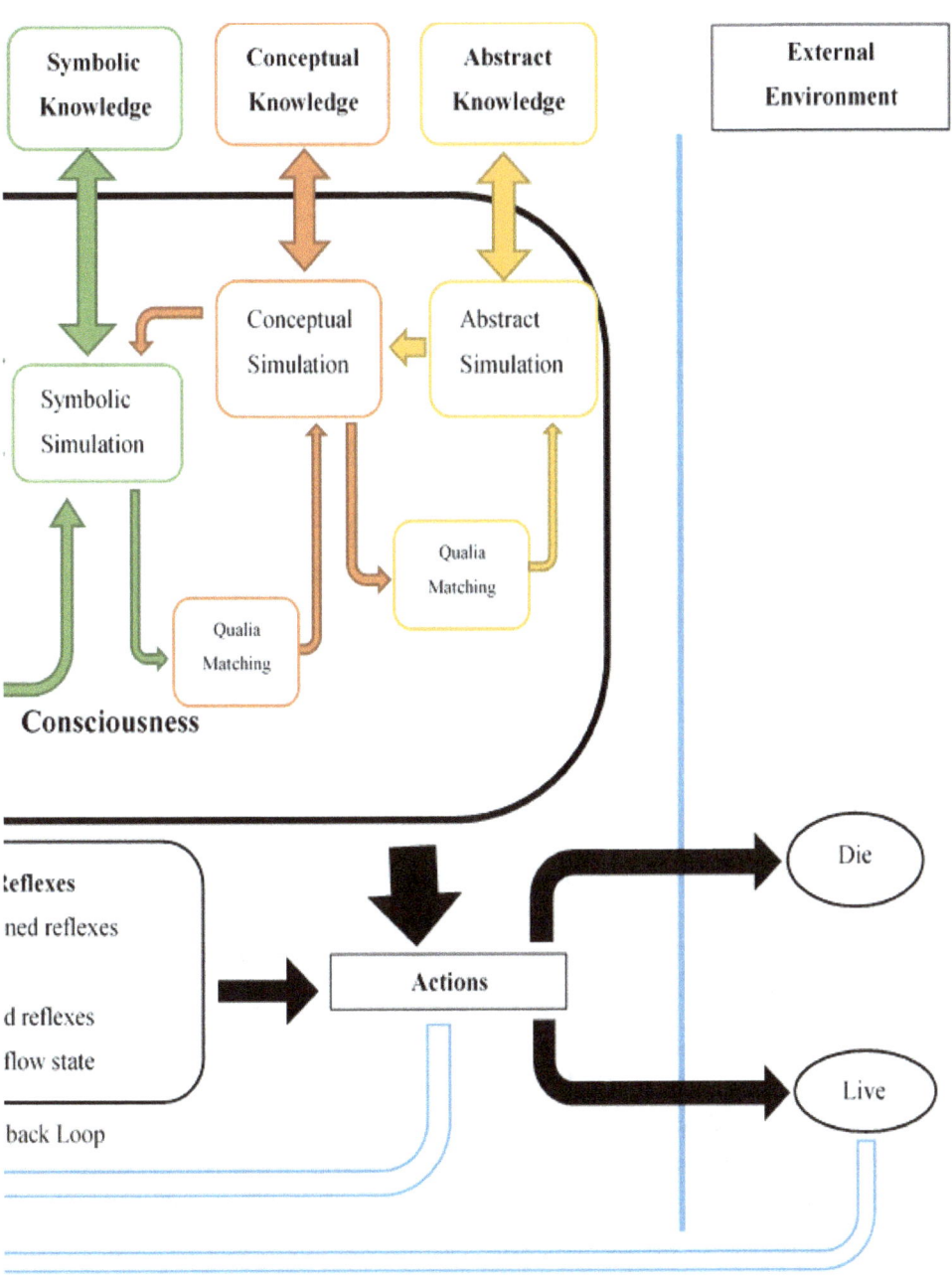

Figure 14 Design of Consciousness expanded Part 2

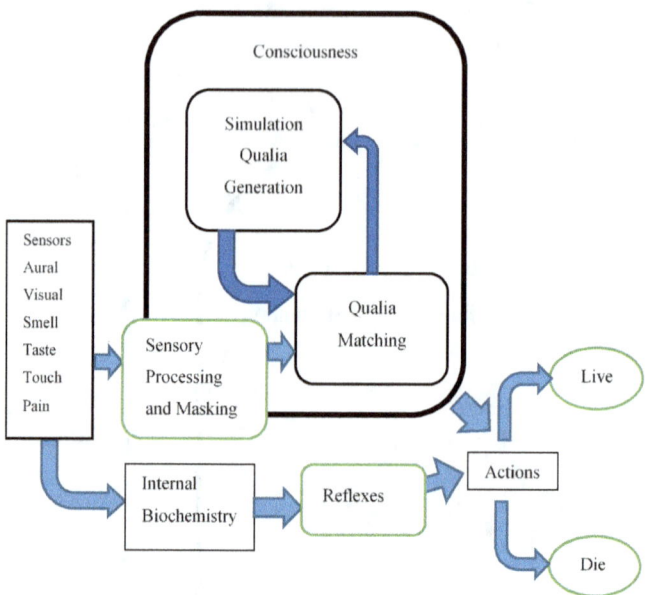

Figure 15 Design of Consciousness simplified

Figure 15 is a simplified version of the system design showing the information flow and general processes involved in the design. We can expand this to include all of the different sensors and their individual processes to be incorporated in the consciousness. To start we examine the system inputs i.e. the sensors, and follow the information through the processes.

There are several information loops from sensors through simulations (both self and world) and knowledge bases. These are related to memory learning and understanding.

The central idea explored in this book is that consciousness is a phenomena that emerges from the simulation feedback loops. This suggests that we feel things because our brain suggests we should be feeling them and these qualia "feel real" as there are real qualia that are combined with the simulation qualia that amplifies any effect.

This might be how human beings and other biological work in principle. It also might not. However, as Alan Turing, discusses in "Computing machinery and intelligence." [64], if a system can imitate another accurately enough that they cannot be distinguished from an external examination they may as well be the same thing. What we can establish is the functions that must be present in a conscious being.

The purpose of the rest of this book is to explore how this idea is similar to what we see in nature and how we can examine it further.

2.1 Sensors

The design starts with the sensory inputs of the consciousness. The inputs include visual, aural (hearing), smell, taste, touch and pain receptors. These inputs can be separates into internal and external sensors. The internal sensors can be combined with a self-simulation and this becomes the idea of self-awareness or self-consciousness.

They type of sensors and the "data" they measure determines what we can become conscious of. They may be simple point sensors which have a specific location, such as eyes or ears. Or they can be distributed sensors like touch or heat sensors.

2.1.1 Internal Sensors

Internal sensor show the state of the entity itself. This includes everything from where our limbs are positioned, how hungry or thirsty we might feel. These senses are often related to our internal chemistry, such as glucose levels, or oxygen. Many of these senses we are not directly aware of, such as our glucose or other hormone levels. Some of them we can be directly aware of: breathing, heart rate, temperature. We may not be able to directly control some of these measurements. For example we can't directly make ourselves warmer just by thinking about it. However, it is possible to be conscious of these internal sensors and make decisions that can optimise them. For example if you are cold, put on warmer clothes.

2.1.2 External Sensors

External sensors are telling us something about the environment we are in. What can we see, hear, smell, taste and, touch.

Human external sensors include our eyes, ears, nose, mouth, and essentially skin (Which allows us to sense things physically by touch). Other animals have variations on these senses with either more or less sensitive or accurate.

Our senses essentially define the playing field our consciousness operates in. We can examine the statement: "are you conscious of the temperature?" In this context conscious means can you feel the temperature, or are you aware of the temperature. In the context of sensors, this translates to can you sense the temperature. It may not be the whole concept of consciousness, but we can certainly say that our sensors are at least part of consciousness.

Conversely, if you can't directly sense something you are probably not conscious of it and probably can't be conscious of it.

2.1.3 Contact sensors

Contact sensors can be used to determine if an external object is touching the sensor and when combined into an array, the location of the contact. Contact sensors can also be expanded to be pressure sensors that can also show how much force is between the external object and sensor. The biological example is skin which can sense contact, pressure and give us a sense of where the contact occurred. In industry we have touchscreens, which can use either resistive or capacitive electrical signals to determine both a contact and its location.

The location sensing of skin in humans can be mapped to allow a general three dimensional perception of where objects interact with our bodies. This is useful for self-perception in helping people build a model for determining the relationship of limbs. This in turn allows people to make complex motions such as walking and grasping objects effectively.

2.1.4 Heat Cold

Heat and cold are very useful for biological creatures to determine if they can survive in an environment. In biology this is typically in nerves in skin. The nerve connections in the skin allow a mapping of heat, cold and pain to the body.

Technology has found many methods to measure temperature including: thermometer (differential pressures with mercury or some other liquid), electrical resistance and, Lasers. These are typically single point sensors, but with some ingenuity it should be possible to create a similar array of sensors.

2.1.5 Chemical – Smell and Taste

Humans have two main chemical sensors the Nose and Tongue. These are used to detect food or spoiled food, mates amongst other things.

Technologically we have Gas Chromatographs which use spectroscopy to determine the chemical composition of a sample.

2.1.6 Sound

The human ear, specifically the Cochlea distinguishes between frequencies in sound waves.

In technology we use microphones that record the soundwaves.

2.1.7 Vision

Cameras and eyes give a two dimensional representation of the world around us.

We typically split images up into three different values red green blue. The combinations of RGB values gives us different colours.

A common question when thinking about consciousness specific to vision is "What is Red", or "What is Blue". The fact is we might have different experiences for each colour. We typically map colours together in something called a colour wheel. We can actually rotate or flip the colour wheel and it would not noticeably change our experience. We would simply re label the colour wheel. What we can say is that the shading between the colours must be consistent, so we cannot randomly change colours around the wheel, swapping say cyan for blue as there is a noticeable relationship between these and other colours.

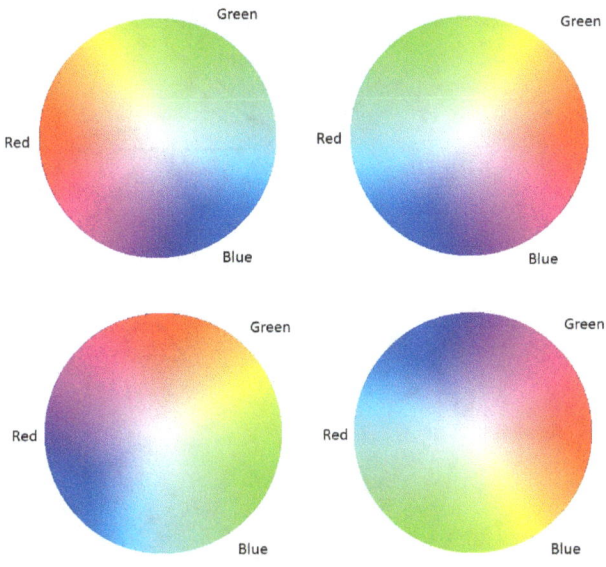

Figure 16 Colour Wheels

Any combination of rotation or flipping will still be valid interpretation as long as the labels remain in the same locations and the sequence between the colours does not change.

2.1.8   *3D Vision*

An easy way to measure the world in three dimensions for a living creature is to use two eyes. With stereo vision matching is performed between the two images and produces areas that are matched together. It is possible for the matching areas between the two images to be used for a model for future matching efforts include tracking the object.

In machine vision, stereo vision requires matching to be performed between the two eyes. Also matching is performed over time to track objects. HPC has been demonstrated to use template based matching to identify numbers and other objects [33], [65].

We can also see that the brain has a structure where the nerves from the eyes swap sides. As in the nerves from the left eye are routed to both Thalamus and finally to the Primary visual cortex. This structure is something that would be required for stereo image processing.

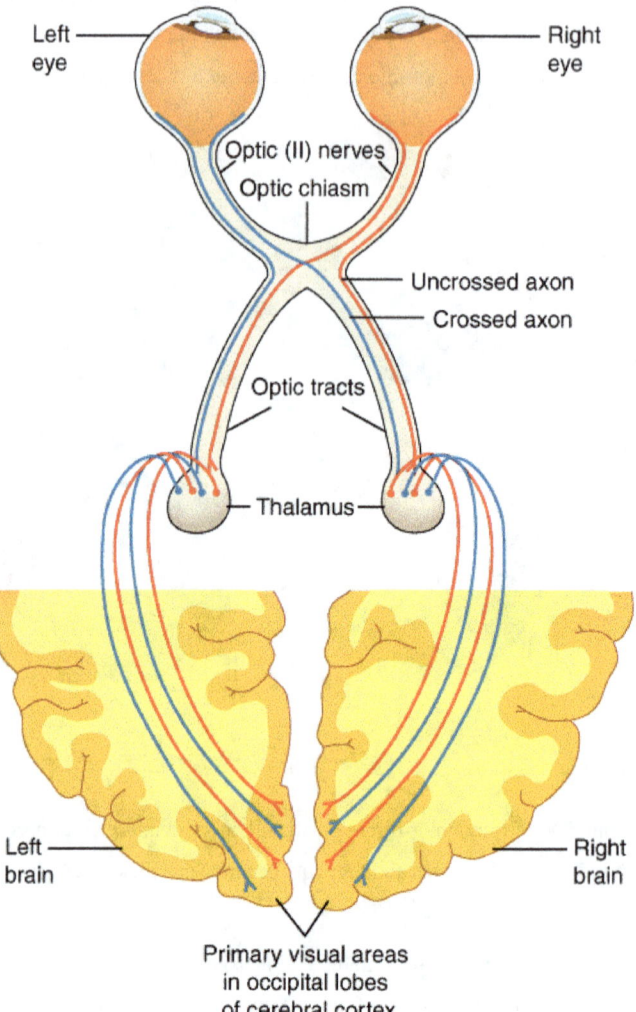

Figure 17 The Primary visual cortex

In industry we have a few more options: Depth Cameras, sometimes called Flash Lidar or Time of Flight (ToF) cameras. These allow depth to be measured directly which allows a 3d model to be generated directly from the camera data. This allows a few processing steps with stereo vision to be skipped and allows a more reliable matching process to be performed.

2.2   *Sensory Processing*

The sensory processing stage has a few tricks to minimise the load required to process the data and allow different areas to be highlighted for specific tasks. This includes focusing and change detection.

2.2.1 Focus – Sensor masking

Do I need glasses??

Being able to see or hear things can be confusing. If too much information is input into a consciousness much of that information can lead to conflicting actions. So what we want to be able to do is ignore some of this information and focus on specific things we can sense.

In terms of signal processing there is a process that is popular. With visual processing we take a source image as such:

Figure 18 Original Test image

Create a mask of an area of the image. The black areas are ignored and the white areas "pass through" the mask.

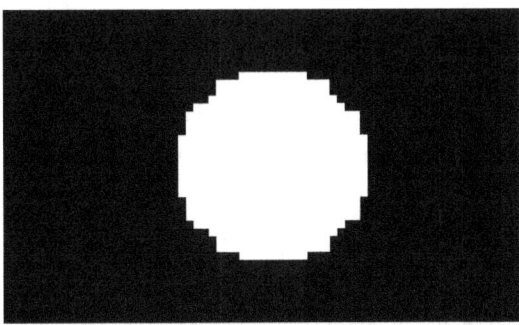

Figure 19 Image mask

This mask is used to invert the relevant area of the original image.

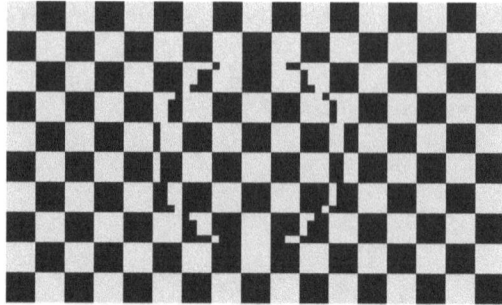

Figure 20 Original image with inverted mask area

Then subtract the inverted image from the original image.

Figure 21 "Focus area" using a masked image

This results in a highlighted focus area and a "dimmed" non-focus area. This technique can be applied to audio signals as well as visual signals. It also has the same effect: dimming background objects and highlighting focus objects.

The focus area is also "more" intense than the original image. This is because the "inverted focus image" is subtracted from the original image. Because the focus area is inverted it effectively adds to the original image making it brighter.

This is similar to the technique used in the design of consciousness. Background objects can be removed and simulation objects can be highlighted. The idea that we focus on some things and this allows us to be aware of specific things and unaware of others. It may also be an explanation of why our senses "feel" real. Essentially, reality "feels" real as our brains are creating a world model that is reinforced by our senses.

This is a simple example of how techniques used in engineering, signal analysis or AI can be applied to the concepts of consciousness, a much more biological or softer science.

The mask demonstrated here could be generated from many different sources. For example:

- Movement from image difference or changes in Auditory information
- Correlation between stereo pairs
- Object recognition, blob detection

The way I think about Focus and this mask is that it covers all sensory inputs and possibly more internal processes. This allows a creature to ignore much of the information they receive from the real world and "focus" on the aspects of the real world that are important to the creature's current goal.

One thing to note is that changes in external inputs when using movement or change detection is that a large "signal" is produced when changes occur in external sensory data.

Have you ever attempted to sleep on a long flight? What I find is that as I am falling asleep I notice that the drone of the engines disappears and this ends up waking me up. The way I understand this is: the mask is suddenly kicking in and filtering out the engine drone and artificially give me a signal that suggests the world has changed and wakes me up.

We can image how this would work in an evolutionary sense. With the predator prey arms race, prey needs to detect predators. So if the prey develop a motion detection using a sequential frame difference mask system it will be able to detect objects that are moving or moving towards it. To counter this Predators need to be able to recognise a target creature and looking for specific object is Object recognition. Computer based object recognition in 2d starts with image segmentation. This means that the image is broken up into potentially interesting areas in the image. There are a few different techniques for this including closed loop edge detection and "blob" detection. At a basic level blob detection essentially finds areas of an image with the same colour. This is illustrated in the sequence of images shown in Figure 22 below.

Figure 22 Blob detection resulting in a mask and isolated face

(https://pxhere.com/en/photo/1569147)

The processing sequence shown in the above images is as follows:
- Original Image
- Posterised/blob detection image
- Single Blob detect plus cleanup
- Original Image with Final blob as a mask

I bring this up as there is some suggestion that the visual processing that occurs in the brain performs blob detection as part of the visual processing parts of the brain (Primary Visual Cortex).

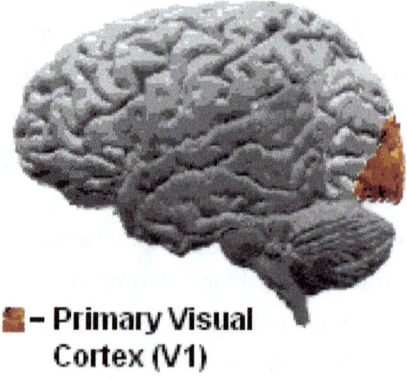

Figure 23 Primary Visual Cortex

So we know this type of processing occurs in the brain, an arguably conscious lifeform, and we know that this type of processing allows an image to restrict image processing to a small area of an input.

Why is "focus" such a big deal? If you have managed to read this far you will remember that a focus is a characteristic of Global workspace theory. As shown by Baars this allows us to reduce the complexity of information we are dealing with and attempt to make sense of it. The mask can be used to highlight areas of sensory input for further processing, for example determining if a predator is moving towards us or identifying other people such as an attractive potential mate.

2.2.2 Thoughts Overriding senses

Another type of masking may occur when we are deep in thought. Often people are described as lost in thought, so busy with internal imagery that eternal stimulus can be ignored. The same mechanism that allows people to focus by masking sensor data could be used to completely mask out external stimulus. This can occur for both visual and auditory stimulus so presumably all of the other senses may have the same mechanism.

2.3 The Self Model

I have a self-model ... Therefore I am! ...?

Much of the research reviewed in this book [15], [17], [19], [21], [26], [48], [50], [51], [66]–[69] is somewhat vague on the concept of what is "self". Much of the previous research uses the psychological idea of self, which is more related to the ideas of our values and attitudes as opposed to a physical interpretation. A few even claim that self is simply a "delusion of common sense". This, I find most unsatisfactory…

One of the fun parts of making a robot is wiring everything up and plugging it in. The problem is when you have a lot of wires you need to figure out which wire connects to which sensor or actuator. If you have a robot with four independently steered wheels, you need to know which wheel is which and which steering joints move which wheels. For that you need a model.

Similarly, as a human being you generally have a good idea of where all your limbs are. Feet at the bottom, legs then, torso, arms at the sides and head at the top. It is useful to know where all of your limbs are especially with respect to your eyes. This way you can determine if something you are looking at is part of your own body.

In robotics you could build a 3d model of the robot and use this internally to map where the limbs and sensors were. This also allows sensors to be mapped onto the 3d model as well. For example mapping touch sensors to a 3d model would allow the robot to know where on its body were in contact with external objects and how much pressure was on each contact. This would include feet touching the ground. This piece of information lets you know if you are standing still, walking, running, jumping etc. This Combined with the position of limbs and visual cues gives the robot a reasonable model of what it was doing.

I would expect the same process occurs in humans and other animals. We do know where our limbs are with respect to each other. As such, I would expect that there is a part of the brain where sensory information and motor control are mapped to. Unfortunately to my knowledge this location has not been discovered. With brain research one of the techniques used is to look at what capabilities are disrupted when brain damage occurs.

There are several notable cases where motor controls are disrupted and people feel like they do not have control of their own limbs (Alien limb syndrome [70]–[75]). People have also reported feelings that their body is actually a boulder. All of these examples may represent cases where peoples' self-model has been damaged or effected in some way to the point where their sensory perceptions are being incorrectly mapped.

This means there is two parts of the self. The actual hardware which stores the data and does the processing that allows us to recognise our "self" and a model of this self. The hardware version of the self is often not conscious. The whole point of reflexes is that they operate without us directly thinking about it. We cannot remember every fact or experience we have had at once, it would make an incoherent mess.

In the software side we have a model of our "self" that is built up over time and experience. This model can be used to predict our own actions, desires and feelings, which allows us to think forward about how we would act or what decisions we would make in a hypothetical situation. This allows us to predict our own behaviour and it allows us to predict what happen in the external world. One feature hear is that the self-model can be wrong! We are not necessarily conscious of how our reflexes will act until they do. In most cases it is not significant, for example we might say our favourite colour is blue when actually it is green! Probably not going to kill you. However, in highly stressful situations, such as in war, our instincts can kick in and cause actions we do not expect.

The idea is that we have our hardware that runs our self and both a physical model and an operational model that contribute to what we could consider the self in psychology.

2.4 Internal Biochemistry

The biochemistry of a life form performs several tasks that simply keep a life form alive, including Processing food, routing signals, muscle twitches for reflexive action and emotions.

Emotions are related to hormones which modify the body's chemistry to perform particular actions or encourage particular actions. When we are angry we release forms of adrenalin to increase heart rate to allow stronger muscle contractions during confrontations. Dopamine plays a central role in learning as a reward system for repeating behaviours. Serotonin and Melatonin are related to the circadian rhythms, sleep cycle. Oxytocin is a bonding hormone that encourages bonding between parents and children and between mates.

Our hormones also effect the growth of many types of tissues in the body. Muscle, various glands brain, etc. Food processing to produce glucose that is used in cells for energy.

Different stages require different hormones to aid the development of features.

This internal biochemistry probably the most basic of our reflexes and the chemistry that allows us to operate in the first place. For example, when we exercise, we release adrenalin that increases heart rate automatically. It is primarily not conscious and in many ways is the basis for our reflexes. That said we are often aware of the results of our biochemistry, when we exercise, we can feel that we heat up. We can actually feel our heart rate.

So in terms of the model from Figure 14, we can see that the internal chemistry is the basis of our reflexes. For simplicity the link between sensors and biochemistry has been included as part of sensors.

2.5   Actions

Actions are essential for life forms to allow them to survive in hostile environments where they encounter predators. The basic necessities for acting allow life forms to find food and mates to reproduce, avoid predators and other hostile environments.

The simplest of life forms such as amoeba make very few actions which can be listed as move eat and reproduce. More complex creatures such as humans need to control limbs (arms and legs) to perform many complex tasks, including walking, communicating, typing on a computer. Controlling these limbs can be a complex mechanical task. In robotics this is often described as inverse kinematics, which is a mathematical method of determining a set of joint angles to achieve a pose that places an end effect or at a particular position and orientation. This is a fairly complicated mathematical tool that still has limitations in how it works. Humans have learned how to perform this instinctively. *{Just scratch your nose, avoid the matrix equations}*

Communication is an essential action for operating in societies. Humans use voice among other actions to interact with each other. These actions have evolved to keep "the life form" alive. We can further examine actions in terms of consciousness. Many actions occur unconsciously, such as keeping your heart beating. We might be able to modify breathing when we concentrate on it, but we have to keep breathing and we have physical changes that occur.

The typical actions we find in life forms include: Move, Eat, Run away from predators, and Find a mate. These actions in general terms have developed to allow the life form to survive reproduce and ultimately allow their offspring to survive. There are many potential actions that life forms can perform and many of these would be relevant to an artificial system.

The difficulty with actions is that often we have conflicting goals. How can we determine which actions to perform when the goals conflict or where the

2.5.1 Movement Detection – Sequential images

Movement detection can be very simple to set up in computer systems. Simply collect two images over time and subtract the first image from the second. This will show any movement in the resulting image by creating highlights near the edges of a moving object. If an object moves laterally in the image highlights are created on either side of the object. Notably One highlight is positive and one highlight is negative. This gives the ability to directly detect the direction of movement. If the object moves towards or away from the observer the highlight can completely surround the object.

Figure 24 Image sequence showing a Subtraction with movement right

Figure 25 Image sequence showing a Subtraction with movement towards observer

These highlights can be even more obvious when considering stereo vision. When matching a left and right image together objects that are lifted out of their background also have a similar area that will not match either side compared to the axis between the two images.

This technique could be used as a method of creating a mask that highlights moving objects. From the previous section this mask can be used to highlight an area of an image for further processing, such as object recognition.

This concept also works in audio or any other sensory input as well. A major benefit of this is that it is possible to highlight changes in any sensory data: images, changes in sounds or even changes in sensations such as pain receptors. All that is required to generate these "movement masks" is a subtractive feedback loop over the top of any sensory input.

The interesting thing is you can see how this works by looking at inverse colour optical illusions like the one shown in Figure 26

Figure 26 Inverted Colour optical illusion

The technique of subtracting sequential images is the most simplistic method of creating a prediction. In other words the prediction is that the sensory information will not change. So when it does change the changes are highlighted and otherwise can be ignored.

More sophisticated predictions can be generated using other techniques, such as feature extraction or building object recognition models.

2.6 Reflexes

Actions that occur unconsciously are often called reflexes. In general these are action that keep us alive. For example flinching when coming into contact with a hot surface.

The Design of Consciousness

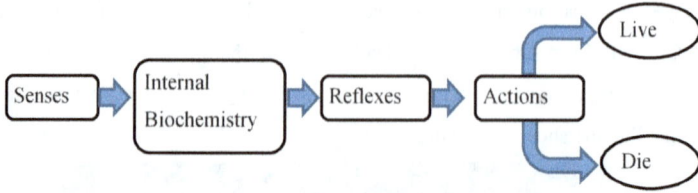

Figure 27 Reflexes

Figure 3 shows a system diagram of how we can think of reflexes. Reflex actions require some sort of sensing mechanism to detect the situation. In the previous example touching the hot surface requires the heat sensors in our skin to determine if the surface is hot. Then we have internal body chemistry to send a signal to the life form's brain where the sensory input can be linked to an appropriate reflex action.

Note the difference between "the design" and the reflex model. Essentially reflexes can be replaced by consciousness to determine which actions to perform.

To decide which action to perform we can look at sensory inputs and directly connect them to output actions, something like what is shown in Figure 28. The action that is chosen can be related to the inputs using a fixed value system (The green dots in Figure 28) where the set of inputs is evaluated in context to each action and the action with the highest value is selected.

This can be implemented using a few different techniques including neural nets/deep learning (Figure 10), Fuzzy logic (Figure 7), expert systems, or hand coding a decision tree or behaviour tree (Figure 6). Either way this allows us to relate sensory inputs and internal biochemistry to actions that may result in different outcomes depending on the situation the lifeform finds itself in and its own internal state. This matrix or system is essentially your personality. The piece of you that determines the majority of your actions.

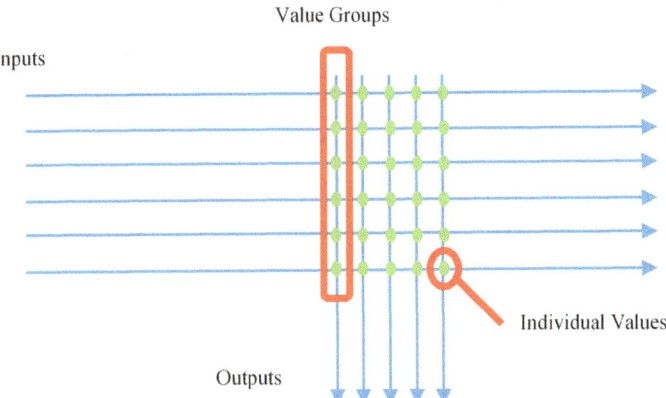

Figure 28 Reflexes to actions detail

The main reason I suspect consciousness evolved is that this reflex system is fairly easy to lock up and fail to produce a single output. What do we do when a given set of inputs results in two output actions to be equally valued? Stand there and do nothing? Probably a bad idea, especially if your life is at stake. What if a higher level of processing was actually used to break the deadlock? The higher level processing: consciousness.

2.6.1　*Emotions*

Most researchers give emotions a lot of interest. Emotions are often considered a central component of mental evolution in biology. The design, you will note, does not include emotion directly. They are deliberately grouped with reflexes and internal biochemistry. I will certainly acknowledge that emotions are a significant mental process. However, this design implies that emotions are not a significant factor in consciousness. I do not know if this is true for biology, but I suspect it is true.

The simulation components of the design can include emotional intelligence either reflexively, or intentionally. An emotional simulation will allow a person to simulate other people and thus create emotional intelligence. The more accurate the simulation the better the person would be able to predict their own and other peoples' emotions.

We would expect that more social creatures, including humans, would be more aware of their own emotions and more sensitive to the emotions expressed by other people. As such we would see behaviours based on emotions and social groups. So biologically we expect emotions to evolve as social structures become more relevant to life forms as they evolve.

We could also develop emotional systems for an AI system in software by tracking various "hormone" or emotional levels and adjusting outputs as appropriate. This is also a terrible idea as we really don't need emotional machines. Do you seriously want your toaster or coffee machine to have an existential crisis when you are trying to get breakfast?

2.6.2 Layers of Reflexes

I discuss much of this later in Chapter 4.3.5 when considering biological reflexes. The distinctions between layers of reflexes made there include:

- Unconditioned reflexes
- Emotions
- Conditioned Reflexes
- Conscious Flow State

The most upper layer of this structure to determine actions would be conscious thought which comes from simulating ourselves and the world around us. My conclusion is that consciousness allows us to make conscious decisions about which actions to perform especially when there are conflicting possibilities and allows us to focus on learning new skills. This idea is summarised in Figure 29.

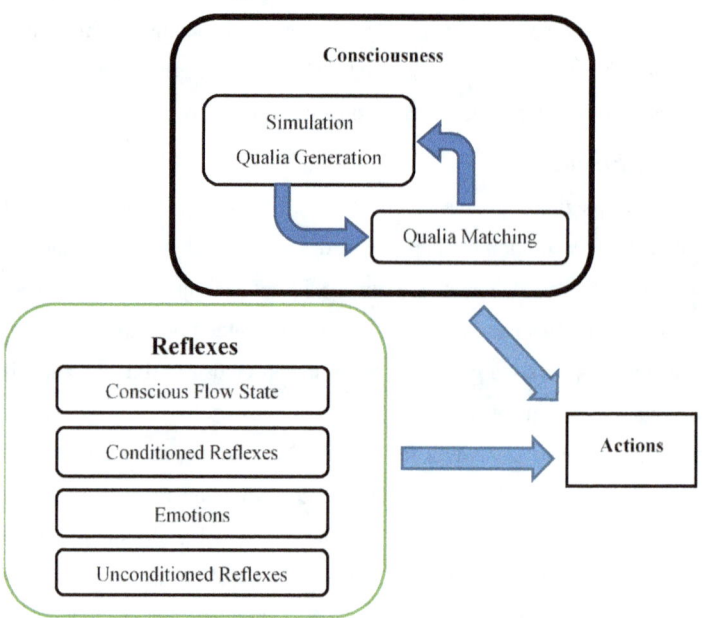

Figure 29 Expansion of Reflexes and Consciousness

In summary I consider any "unconscious" action a reflex in this book, mainly to allow the different levels to be grouped into a single structure to simplify the design. This may be a good area to develop in future work as these "reflexes" are still interesting in themselves, however they are simply not the focus of the research presented in this book.

2.7 Perception

I recall looking for a screw I dropped while fixing something at home. I noticed that I did not actually see the screw for a few seconds even though when I did spot it had been in my field of view all along. It was stuck behind a panel but perfectly visible. My guess is that I did not actually "perceive" the screw until it had been through the sensory processing and qualia matching processes.

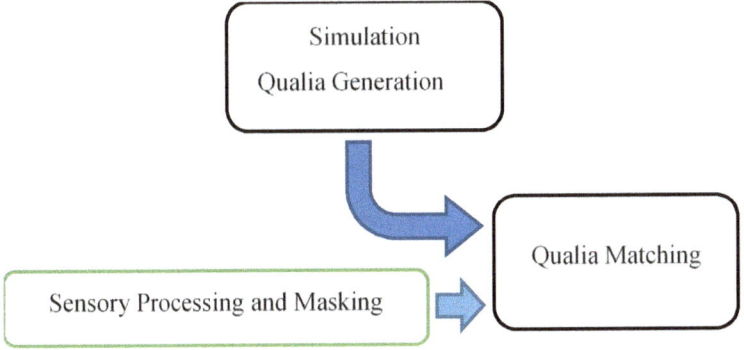

Figure 30 Perception stages of consciousness

My hypothesis is that perceiving the screw would require a template stored in memory to be "loaded" into working memory and matched to the sensory input from the eyes. I will discuss this more in Chapter 4.3.9 comparing between biological and computation AI structures. I suspect the wiring in biology is somewhat more complicated, than what we would consider loading in computer terms, more like connecting to an appropriate program in the cerebral cortex.

2.7.1 Pattern recognition

A pattern is a very general term for a signal in any type of dataset that can be recognised. There are various means for detecting patterns including Feature detection (Harris, Roberts, Sobel, SIFT, SURF, RIFT and many other acronyms). Edge detection (Detecting lines rather than individual locations), Template matching and at the most complex Object recognition.

In mathematics and computer science they are used to find common various types of relationships in data. Computer science has applied these technique to Audio to detect speech and visual or three dimensional data to perform object recognition.

In biology the equivalent recognition systems continuously analyse the world around us to help us identify and interact with objects or people that we know. As we grow up from children we build up the equivalent of a database of objects along with associated capabilities of such objects.

2.7.2 Pattern reproduction

Pattern replication is almost the opposite direction of pattern recognition as in it takes known patterns and uses them to reproduce an image or in, more general terms, a signal.

In many ways this is in the same conceptual space as Generative Adversarial Networks (GANs). These are algorithms that produce patterned outputs. To create a GAN we can connect the output of the pattern reproduction to a pattern recognition system and determine if the produces a stable loop, in that the connected system produces an output that it can correctly interpret as the pattern. This feedback loop would create a stable but adaptive system that would continuously recognise any initial object fed into the system. The idea is that this feedback loop effectively becomes working memory in biology.

2.8 Simulation

How do "I" define a simulation?

My Background is in Mechatronic engineering, one central topic of which is manufacturing systems. Part of what robotics in manufacturing requires with the complexity and capabilities is to determine if the design will actually work correctly. My work was to put a 3d scanner on the manufacturing system to be able to detect unforeseen events. I realised that by comparing the simulation to the real world this could be defined as the system being aware, one of our keywords from the introduction. If we have a model of what to expect and we can update the simulation, we can extrapolate and make a prediction about what will happen next. This sounds to me like understanding, another keyword. I was using a mechanical system and making predictions using the laws of physics, the capabilities of the robotics simulation system, and inputs from a real robot. The key point is that a simulation is active and makes a prediction about what might occur.

The next question for myself was: can this concept be extended to encapsulate other types of situations? Can we "simulate" behaviours of life-forms or people, society? ... Probably. To expand what we mean by a simulation we can think about people: can we simulate auditory phenomenon, animal behaviour, a person's feelings/emotions, or a social situation? If we can simulate these kind of things in a computer could we be doing the same thing in a human brain?

So lots of questions there, but back to simulations.

There are many examples of simulations in the computing worlds, everything from calculating weather patterns, material physics, to computer games. Many of these again use what we know of physics to make predictions about the real world.

Computer games again have a huge variety of types of games which look at pattern matching with something like Candy crush, or Tetris; we can simulate the construction and operation of entire cities like SimCity, Skylines, Cities XL; we have social simulations like "The Sims"; just about anything we can think of we can come up with some kind of simulation for.

To sum up, the way they are used a Simulation is a general term that encapsulates any process that is used to make a prediction. The features include the use of existing knowledge about concepts in the world and how these change when they interact to make a prediction.

Can a human being make the same kind of predictions? Not with the same precision, but humans are good at different types of things, social situations, interpreting speech etc.

When we talk to ourselves we can literally hear our own voice in our head. So in terms of simulation this suggests that we as human beings are capable of generating or "simulating" human speech mentally.

Social situations require us to look at other people and from their expression estimate their state of mind, i.e. are they happy and can I interact safely or are they angry and going to attack. How do you figure that out? You simulate the person in your head and based on the expression you simulate how you would feel making the same expression.

What this shows is that there are many types of simulations that can be performed and that humans specifically can perform some sort of simulation ourselves to predict various outcomes in different fields. This could be constructed in several different ways. In terms of the design and humans there may be several simulations running concurrently. There are at least three types of simulations I can think of: a self-simulation, a world simulation and many types of symbolic simulations.

The most basic simulation we can consider is a simulation of the incoming sensory information. We can take templates stored in memory of features in a sensor feed and generate an image or audio feed that matches with the raw sensor data. From this new generated image we can also perform object recognition. This cycle will become stable if the same templates used to generate the image as the ones that are detected from object recognition. So the simulation in this context is the ability to generate a sensor image from templates or known objects.

If I was developing a computer program the self-simulation would be used to map the inputs from the various sensors into a coherent model of the "self-simulation". Literally, a 3d model of the physical layout of the inputs and outputs. Add in the dynamics and planning system and we have something similar to a 3d animation system using inverse kinematics to estimate the required position of limbs. Using an interpolative system it would be possible to create "animations" and with a control system these animations could be translated to control the movements of a real robot. The self-simulation then allows us to predict our movements and how we can interact with objects in our environment.

The world model is used to map the surroundings and the place in it. Part of this process what modern robotics called "Simultaneous Localisation And Mapping" or SLAM. A working knowledge of physics and a world model allows us to predict events that occur around us. This also allows us to interact with the world. One task requiring a world model to exist is navigation. We need to know the relationship between ourselves and objects in the world to be able to interact with them.

The third type of models I call symbolic. These are higher level simulations that model symbolism, like words, concepts, social situations, or other people. These simulations allow us to simulate the more complex situations and allow us to make more complex predictions. There may be several types of simulations that operate at the same time, but in different stages in the mind (conceptual and abstract). The self-model can be separated into a few specific types. Processing visual information can use several simple forms of a simulation. The simplest is to assume that optical information will remain constant. By subtracting the previous frame in video processing isolates out movement between frames.

A simulation can use symbolic information such as known objects. For example, these can be reprojected and matched onto an image. In operation this becomes a recognition problem. In other words: How do we match symbols such as letters, in images, words in sound with known symbols stored in <u>long-term</u> memories?

These "micro" simulations can be used to estimate the inputs from a sensor system (Audio, Visual, Touch). This creates a feedback loop that allows the differences between the simulation and sensory input to be minimised. Other simulations include Auditory, "mechanical" output, Emotional, or even Social simulation.

If you have read section 1.3 carefully you will notice a similar operation between Hierarchical Predictive Coding (HPC) and the simulation feedback loop. One of the main features of HPC, especially with optical processing, is that feedback loops are used to minimise the error between the sensor input and the internally generated signals.

The simulation also has many features in common with Global workspace theory (GWT). GWS shows the mind using local memory to help focus on what we are interested in. The simulation system must also use memory to have active elements in it. As we look at an environment we build up a model of the environment. This environment is often broken down into smaller subsets, objects or visual phenomena (lights, reflections, shadows etc.) that we recognise. The objects we recognise must be in memory as part of the simulation.

This demonstrates that this design at least has the potential to demonstrate many of the same features as found in other research such as HPC and GWT. As such I think there might be a common structure to what we feel is consciousness in lifeforms and what we can define as consciousness in the robotics and computer systems we as the human race can construct. This structure I am referring to is "the design of consciousness". By comparing expected outcomes from a simulation with the real worlds might result in a process that if looked at epistemically (if you were to experience it directly) would resemble consciousness.

2.8.1 *Self-Simulation*

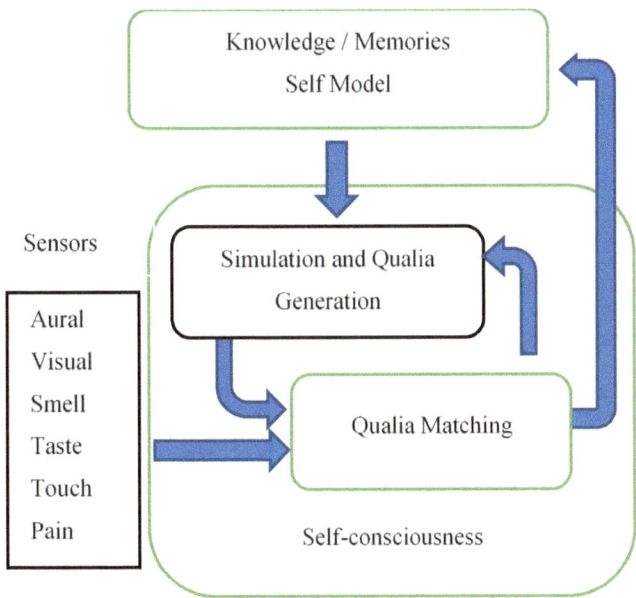

Figure 31 Detail of Self Simulation

In terms of consciousness you can think of Self simulation in terms of your own body. We all have a 'simulation' of our own body which you can test yourself.

Please read the whole instructions before you do this as you need to close your eyes. *{And don't poke yourself in the eye... It hurts when you do that.}*

- So close your eyes and think about your hand and your nose.
- Point your index finger and place it on your nose.

You should find that you will get close and if you have tried it a few times you will be able to do this.

If you think closely about this you will realise that you will have an understanding of both the position of your finger and your nose. In mathematical and robotics terms this action can be completed using inverse kinematics or forward kinematics. Both of these techniques find a mathematical solution for all of your joints that add up to putting your arm joints in the correct positions that your finger is on your nose. Mathematically, this is a fairly complicated task and gets more complicated the more joints you have, but can be solved in most cases.

The brain has two areas that map inputs, the primary sensor cortex, and outputs the primary motor cortex. In robotics you have the same problem all of the wires that route from sensors to the central controller and back out to any motors or actuators need to be mapped. The computer itself only has the wire ends, so another mechanism is needed to map where these inputs and outputs are in relationship to each other. Internally in a robot, a 3d model could be used to map out where all of the joints and sensors actually are. I suspect that something similar exists in the brain.

The point with consciousness is that you are conscious of the position of your hand and nose. In terms of robotics a reference frame hierarchy is used to represent the joints with rotations and offsets so accumulatively they are able to determine physical relationships, distances and angles between your hand and your nose. So I would expect that somewhere in our mind we have a model of our own body. The second part is that with a simulation we also know where to expect our bodies to be. This gets to the idea that part of consciousness or awareness is as much the comparison between what we expect from the simulation and the signals we receive from the real world as it is these parts independently.

We can see this kind of thing happening in nature fairly regularly. Puppies will chase their own tail as they grow. Eventually they get hold of their tail and bite it. When that happens they will be able to connect the pain they feel from their bite to the visual stimulus when they see their tail. This shows a learning process that updated the understanding of the puppies own body.

We can also do this with robotics. My previous research examined real-time 3d scans of a real environment along with a simulation of the same environment. The simulation used industrial robotics software (ABB's Robot studio) that also ran the real robotic arm. The theory developed showed that the 3d scan could be matched to the simulation while the robot was running. From the matching algorithm objects from the simulation scene could be identified as being present in the real world scene as shown in Figure 32 Robotic simulation and Scanning.

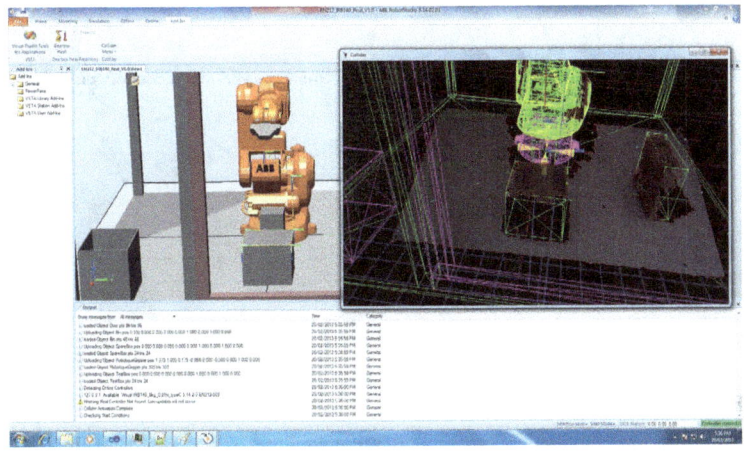

Figure 32 Robotic simulation and Scanning

As mentioned previously, this research became the seed for the idea of consciousness and the basis for the design of consciousness that was developed. The central idea of this design is that we simulate ourselves. We literally have an internal 3d representation of ourselves that we can match to incoming sensory data. The self-model is Ontologically Epistemic, your physical self exists in the real world but its operation can only be viewed internally.

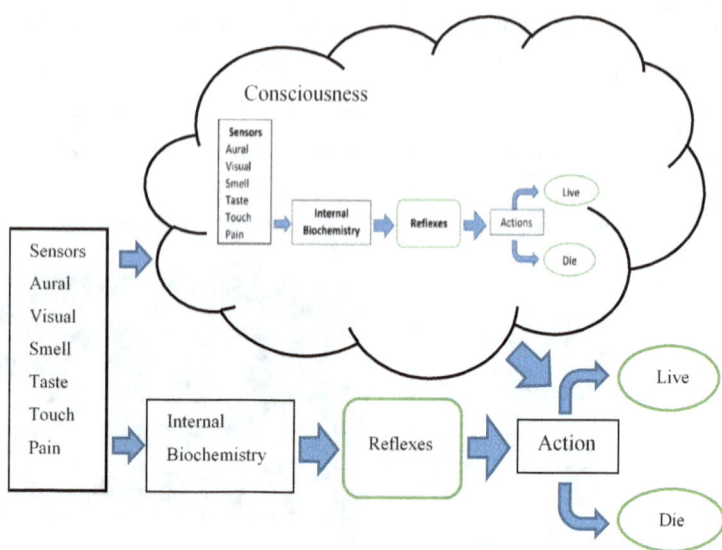

Figure 33 Self Simulation and Consciousness

From Figure 33 the internal operation of consciousness we in effect use the self-model to predict our own actions. This is useful for testing what actions we should take. Sometimes you get it wrong! This becomes a learning opportunity. For our self-model this is what we do when we grow up from being a baby. We need to figure out where to place our feet and adjust or weight to be able to walk. The first few tries are probably not going to work well. With practice our muscles get stronger and we update how well our arms and legs work. Eventually, we figure out how to walk *{or run and fall over}*.

The idea of a self-model is easily expanded to include the world. We can see and recognise our own hands, so we can see if we are holding something and potentially recognise this object. The object is definitely external to our self. That said the same strategy we use internally can be used to interact with external objects and understand how it operates. Essentially, we build a model of the world.

2.8.2 External simulation: The World model

Being able to understand our own actions allows us to figure out how these actions might occur in the rest of the world. From our experiences in the world we build up a model of what is in the real world and how it works.

As we move around the real world and interact with it we build up an understanding of the world. My interpretation of this is that we build up a simulation of how the world works as shown in Figure 34. We learn rules about objects, spatial relationships, relationships between actions and effects to do with objects. This can be described as a simulation when we consider these rules actively operating. For example once we observe an object being knocked off a table we observe it falling off and breaking on the ground. This becomes a rule for us: when we knock something of the table, it breaks. Hence in your brain you may realise that if you too many objects on a table, you may realise that one of them will fall off. The simulation is an active understanding of what does and will happen in the future.

Because we can use a simulation to make a prediction about what will happen we can determine if our simulation is correct by comparing what happens to our expectations.

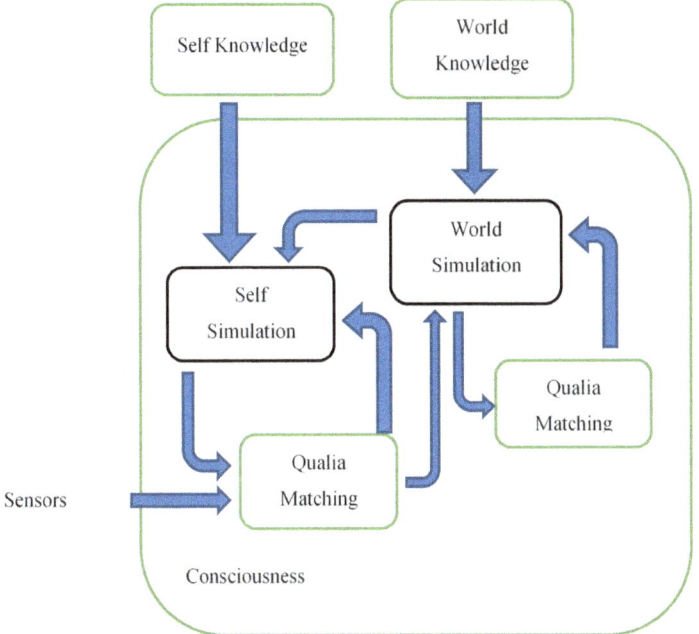

Figure 34 World model system diagram

2.8.3   *Symbolic Analysis*

Symbolic analysis takes chunks of sensory data and looks for known "symbols" to identify. Most likely the focus, change and blob detection in visual stimulus are used to identify particular regions of an image for further processing. These blobs can the examined using object recognition for known visual items or even things like actual symbols "letters" words signs etc. The same thing occurs with auditory systems. We break down the sounds we hear into phonemes that are then reconstructed into syllables, words and sentences. Baars [19] gives a good example of how auditory stimulus is analysed by specific processors into understandable language.

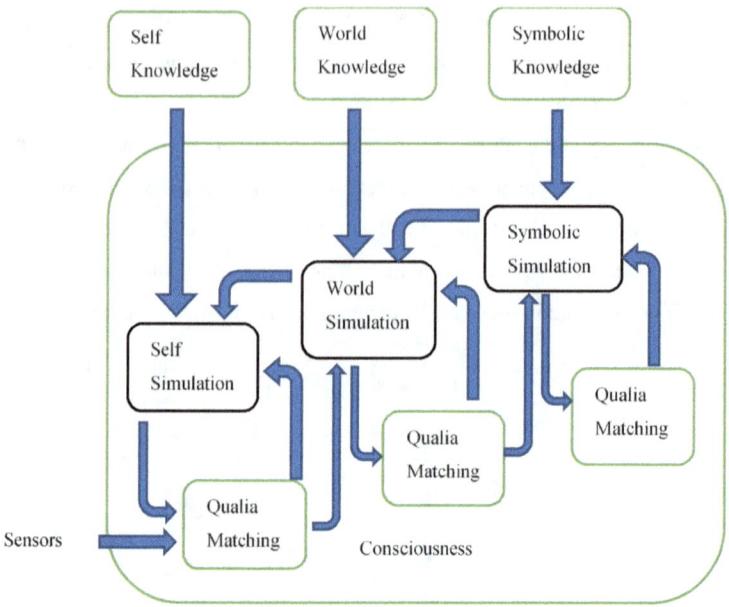

Figure 35 Symbolic analysis system diagram

From Figure 35 we see this looping structure emerge, where we have internal models that allow us to predict outputs from simulations and match this data to incoming sensory stimulus. This again starts to look more like the structures identified with HPC.

2.8.4   Conceptual Analysis

From symbolic operations we can take it another step forward to examine a conceptual simulation. The idea of a "Concept" we are defining hear allows symbolic data from multiple sensory systems, the most common being visual and auditory stimulus to be combined into a single entity, a "Concept".

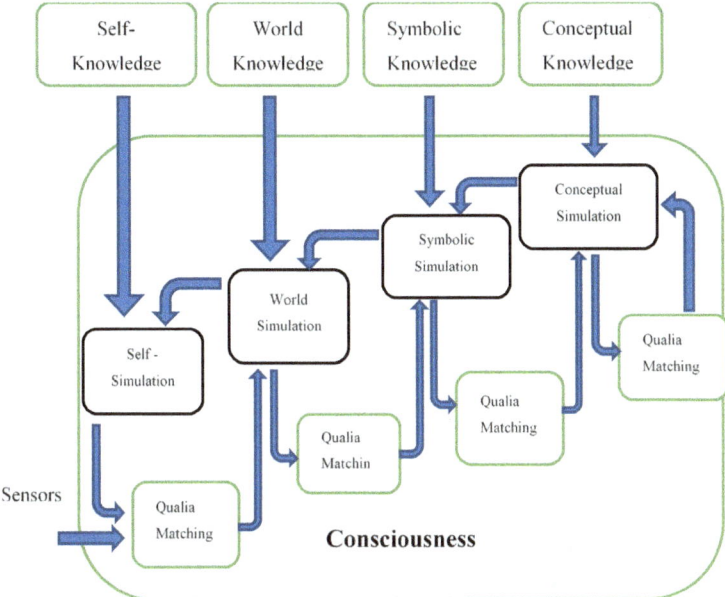

Figure 36 Conceptual analysis system diagram

We know that in the real world there are objects and these objects have many properties. If we take a dog for example we can look at many different properties. We can see them including their variations. The make noises the most common of which we call a "bark". The Concept allows us to group the visual and aural stimulus under a single classification as shown in Figure 37. We can also develop a programming structure called a class that could group all of these ideas together, shown in Figure 38. This is a very simple self-recursive structure. Technically this is not the best programming practice, but it allows us to see how we can use concepts to represent everything in our own minds. This structure shown that an image is a concept. The image itself has properties: size, features colour. These ideas can also be represented as concepts.

The idea presented by Edelman that concepts may be represented without a name at least in animals. I suspect that each concept requires some way to access it. As we have language I have used names or words as accessors. Essentially each concept would have some way to find the word and look it up, like an index for a book. These accessors would be in the same "language" as the sense. So a word could be a visual accessor or an aural one i.e. an image or a sound.

There are two types of concepts we can consider, especially as a part of a simulation: Properties and Operations. Properties describe the concept itself. Operations describe what the concept can do, much like in programming with methods and attributes.

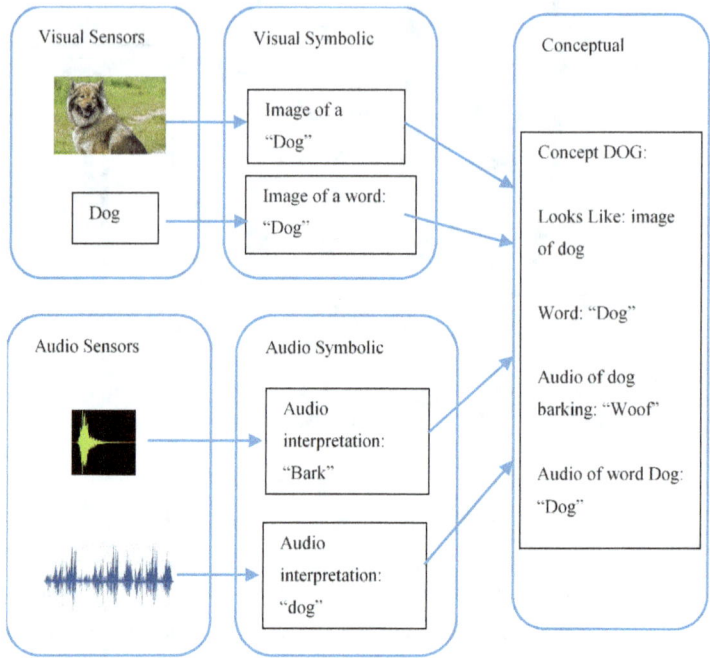

Figure 37 Conceptual analysis operation

```
Class Concept
{
public:
        Concept() //basic constructor
        private:
        string m_Name;
        string m_Decription;
        map<string, Concept*> m_Properties;
        map<string, Concept*> m_Operations;
};
```

Figure 38 A basic class structure

This structure becomes quite flexible we can use it to describe pretty much anything we find in the world. Most importantly we can use this structure to describe how objects interact. This is extremely relevant to simulations as a simulation is active. We can do stuff with objects, and some objects (especially animals, people etc.) are active. They do stuff without having to interact with them.

The operations are key to the simulation becoming active. We know that gravity exists and makes objects fall "down". When we let go of an object we know that you might be able to catch it if you get your hand underneath it quickly enough. There are many other operations that can be performed on objects, throwing catching, painting, etc.

If we can determine the rules of the world accurately our internal simulations can be useful at making predictions about the world. This gets into determinism which I will discuss in Section 7.3 in more detail. The more deterministic the world the easier it is to determine the rules of the world, and we have found a lot of them (Physics, Biology, Mechanics etc.). When we decode these rules, the easier it is to make predictions based on them that can be accurate.

This shows is that the more deterministic the world is the easier it is to make accurate predictions. The more we know what will happen the more we can turn around and say no this is not what I want and attempt to change the outcome. In other words the more deterministic the world the easier it is to exercise free will.

In terms of predictions the final level we can consider is abstract concepts.

2.8.5   *Abstract Analysis*

From conceptual we might have an even higher layer of processing for completely abstract concepts or this may simply be a part of conceptual analysis. I am not sure if this would require a different mechanism for a human mind to work with, but with the Concept as defined in the previous section it should be possible to generate an Abstract concept with the same structure.

So, what is the difference between a concept and an abstract concept? The idea is that an abstract concept has no real world instantiation. Mathematics is a good example. There is no real world physical object for the number zero. Yet we can still represent it in the form of a visual image "0" or as a spoken word.

Abstract Concepts are often difficult to understand in terms of what they are as we always need to have a way to reference them, a name. The name needs to be something we can see or hear and identify the concept we are talking about. Again we can extend the design of consciousness to incorporate Abstract Concepts as shown in Figure 39. Essentially Abstract concepts fit above conceptual simulations.

The Design of Consciousness

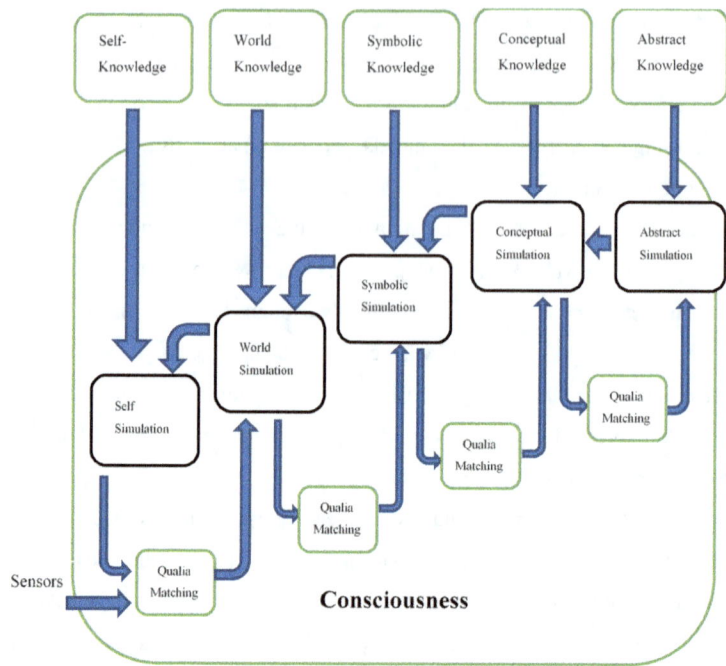

Figure 39 Abstract analysis operation

So what can Abstract concepts do for us? Well high level mathematics is a good example. These concepts are about as abstract as we can get. Geometry is probably the least abstract as it is easy to represent these concepts in the real world. From Calculus we have Differentiation and integration, Statistics gives us mean, standard deviation, confidence and error propagation. But this is all the basic stuff.

What can we do with mathematics? This has allowed us to make extremely accurate predictions. A few notable ones were from Albert Einstein and the most accurate prediction was made in Quantum electrodynamics. Albert Einstein who predicted that time would slow down the faster we moved. This was confirmed some 80 years after he made the first prediction, by flying a plane around the world carrying some extremely accurate atomic clocks. Einstein also predicted that gravitational waves would occur and could be detected if two large objects such as black holes, collided. This was confirmed nearly 100 years after the theory was presented. Quantum electrodynamics predicted the electromagnetic fine structure constant to within 10 parts in a billion (10^{-8}) compared to experimental results. These kind of predictions are only possible due to the accuracy of the mathematics used to describe the models of the real world.

{If only we could remember abstract concepts easily...}

2.9   Memory

After reading a few chapters of Baars book "On Consciousness" I wanted to write down a few ideas I had on memory, the problem was by the time I had gotten over to write them down I couldn't remember what I wanted to write down about memory.
...This struck me as funny.

Information can be "encoded" in writing, symbols, images, electrons, bits and even neurons. It can only be decoded using the same type of system that did the encoding. In our brains the information stored are memories. We remember objects, people, faces, our culture *{Dawkins Zombie: Meme-ory. Cultural memory}*, history and our personal physical and emotional feelings. We also remember being conscious. We can usually remember that we were thinking and what things we were thinking about.

We use memory constantly so it has been studied fairly rigorously in the cognitive and neuro-sciences. Figure 40 shows a diagram of the process of memory acquisition. Any sensory information is encoded in some manner, in our brains this occurs chemically as links between neurons via synapses. So when we receive sensory information it can be encoded into neurons. To access it we need the neurons to reconstruct the input signals to be able to output them again. Ideally memory would be "lossless", but from our own memory know that our memories are far from ideal. My guess is that we only reconstruct enough of our memories to be able to reactivate the concepts we recognise in the sensory data in the first place.

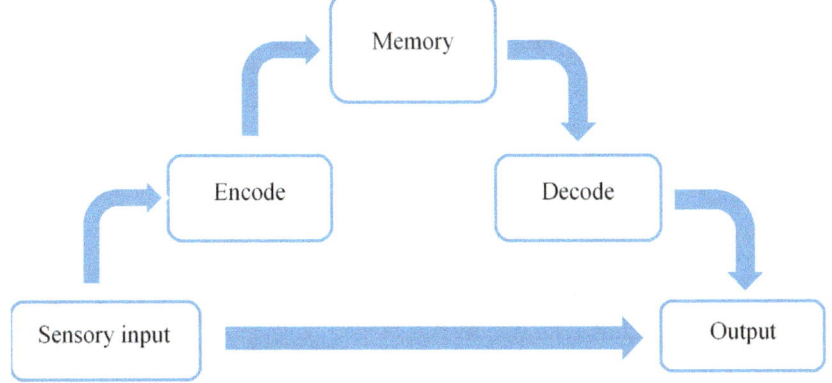

Figure 40 Memory access

One of the main features we have in terms of Short term or Working memory is that we can only think of "7 plus or minus 2" unrelated items. When considering more complicated concepts this can be reduced to 4 items. Why this occurs is still something of a mystery to the research community.

My current hypothesis is that we use what we have in working memory as part of a simulation. More importantly we need to be able to determine any interactions between these concepts we have in memory.

Taking simple mathematical concepts 2+3 = 5. We need to think of at least 3 things: "2", the operator +, and "3". These things then combine to determine the result "5". This means we need to have space for at least 3 items and the relationships for "+" and possibly "=". The point here is that using active knowledge we get interactions between the items we have in memory. The more items we keep in memory the more relationships we need to manage. If we only consider operators with 2 inputs from 4 items in memory we end up with 12 possible interactions. From Combinatorial mathematics this would be four Choose two ($^4C_2$). This also called Quadratic complexity in Complexity analysis, in Big "O" notation: $O(n^2)$. This means the more items in the system the more interactions we need to be able to consider. Quadratic complexity gives the possible interactions in Table 1.

Table 1 Quadratic Complexity

Items	2	3	4	5	6	7	8
$^nC_2$	1	3	12	20	30	43	56

Worst case scenario we need to be able to consider the interaction of every possible group of items, from each item individually to how the items operate as a group and every possible combination in between. This results in what is called factorial complexity (n!). With 4 items we end up with 24 possible interactions. Each interaction may require processing by the brain.

Table 2 Factorial Complexity

Items	2	3	4	5	6	7	8
n!	2	6	24	120	720	5,040	40,320

As you can see from Table 2 with Factorial Complexity this becomes very large very quickly. Even with 5 items in memory dealing with 120 possible interactions is likely to be far more processing than most people, let alone other creature. This may be a reason working memory rarely goes above 4 items when dealing with complexity. Basically 2 input items an operation and a result. The good news is that most of the time the interactions are non-existent or at least very simple, so we probably need to be able to access a few more items to manage sequencing which gets us to at least 6 items and gives us a bit of a buffer with 7 items. This demonstrates that working memory is as large as it needs to be to have simple interactions without significantly draining processing resources and hence energy from a living being.

In terms of computing and robotics memories can be created by simply saving input sensory states to a hard drive. These can be as simple as log files giving a text description of events or as dense continuous sensor feeds dumped to file. Biology is a much more complicated system. Caruthers, p78 [20], Mayes and Roberts 2001 [76] have suggested that memories are stored very close to the area where they are initially processed. There are also parts of the brain that seem specialise in memory access, the temporal lobe, Right sided Posterior Neocortical Regions for memory retrieval [77]. In Computing a technique for managing data is to create an "index", which maps the location of files to various metadata including date-time, file size, file name etc. It is possible that the same techniques could be applied in biology. In other words the memories are actually stored where the processing occurs but are indexed in even a number of locations. For example the temporal lobe *may* be where date-time index is stored or managed. I have no idea if this is true, but I can say that it would work as this is exactly how computer systems currently work.

Following this idea through, human memories are often described as relational meaning that we often recall specifics through finding similarities in connections, such as recalling memories from a particular scent. For example trying to remember when an event happened you might think that it was summer and how warm you felt. Something about warmth will trigger another memory of cooking fresh apple pies. Then suddenly: "It was just after we harvested the apples that Saturday when school finished." This example suggests that we might be able to access different memories using different feelings or in the computing analogy: different metadata.

It is interesting thinking of consciousness and memory, as memories often seem to appear without effort. Considering how files and databases work it does not surprise me that we do not consciously drag memories from storage. A computer system can have thousands of files on a single computer and considering the internet millions or Billions of files with a distributed network system. To be able to access one of those "memories" or files requires fairly complex search algorithms on computers. Biologically we are implemented as a neural network which creates some advantages and difficulties in implementing a distributed "file" access system. As mentioned previously the relational nature of human memories mean we have a tendency to bounce around almost feeling our way through memories.

Interruptions in memory occur fairly frequently. The Tip of Tongue state described by Caruthers[20]. This shows that our memory can fails us. In computing terms the TOT state could occur when we search for "data" from a database and receive no results despite the being sure we have some relevant data. This might occur as the wrong accessor was used. In database SQL terms the SELECT statement WHERE condition is too restrictive. Damage to the brain can is also associated with memory loss. T

The hypothesis developed here is that consciousness requires a combination of predictions and external senses in a feedback loop to be conscious. Given this it would require us to be able to predict which memory is being sought. Obviously difficult before the memory has been accessed. So it would often seem that memories or other thoughts just appear.

The "feeling" of memories is often difficult to describe. One Idea I have had is that human memories could or should include all of the sensations associated at the time of the memory, including the feelings of our own bodies, warmth or cold, pressure, our emotional state, happy, sad, angry. In more general terms: "the experience" of a memory. This would allow the associations to trigger memories as previously mentioned the relational nature of memories. This would also be useful when considering evolution and survival as we would be more likely to remember events that we found scary such as being attacked by a predator.

The hypothesis I primarily work with in this book is that or brains produce or reproduce these sensations then recognise them as part of a feedback loop, similar to Carruthers description [20]. In terms of memories this means we are capable of recalling the sensations we have previously felt and effectively re-sensing them.

In terms of the design there are at least three types of memory. The first is related to ourselves so our self-memory. This would include our physical state, our mental state and emotional state. The next level up would be related to the real world. Things like objects people, names and attributes of such objects. The highest level of memory is related to concepts we build up from the real world. Then there are abstract concepts such as math, honour, duty, philosophy etc. which might be stored as concepts. Things that are not necessarily directly in the real world, but are represented by words or artificial images, symbols.

From the design each level of memory is associated with a consciousness loop: self, world and abstract. This allows us to load previously encountered concepts into working memory and allows the conceptual operations we know of to act on any object we have identified. Effectively memory is used to load known templates into the simulation. We need to know what the objects are and how they interact. Again, the limit in memory (7 +/- 2 items) might be explained by the number of interactions possibly between objects. The interactions are essentially things we can make predictions about from active simulations of the objects in memory.

2.10 Temporal memory

In case you were wondering: Temporal = time

The temporal nature of memory is that we have one point of view with a series of events in chronological order. As such, we would expect that there is some brain structure that allows us to manage the sequencing of our memories and in some cases our actions. Thus we can introduce the temporal lobe in our brains. The temporal lobe is actually associated with language and memories, both of which require chronological ordering. Think about language for a moment especially listening to language.

We hear sounds sequentially. We collect one sound after another until we have a sequence that matches a work or sentence we understand. It's actually fairly easy to implement this concept in computers and is called voice recognition in computer science. We implement something called a que where the sound is recorded in sequence. From here we can break down the sequence into specific sounds called phonemes. These are often described as the perceptually distinct sounds that humans can make. From there we attempt to build up combinations of these phonemes into words, combining words into sentences, then analysing the whole thing to be able to understand what people are saying.

This is often used to perform speech to text for writing and can be used to verbally control computers or other machines. Speech recognition is fairly widely used in the world today. It is still not perfect, but we have a reasonable understanding of how to do this. In biology obviously humans can perform speech recognition and most likely a similar process is being performed using a neural network. Clearly biology is capable of performing this task effectively, but I would expect that decoding the neural network would reveal a process similar to speech recognition in computer science.

The important thing here is that to understanding words we need to keep a chronological sequence. This sequencing mechanism is also applicable to memories, planning and probably many other features of signal processing.

I suspect that the temporal lobe or some closely associated region helps us recognise the passing of time and help with sequencing. In computers there is a thing called a clock-signal. If you look at a computer CPU it will typically say a 2.4 GHz. This means that the processor has a clock signal that runs at 2.4 Billion cycles per second. With human brains we record signals between 5 and 100 Hz and in visual processing a 40 Hz signal is often found.

So how does this relate to consciousness?

If our sensory loop operates using a clock signal to help synchronise sensory input and allow us to match for example visual and aural senses (vision and sound). This allows us to match the sounds people hear to the facial expressions people make while talking.

The clock signal is not intentionally a synchronisation mechanism it literally allows a sequence of commands to be performed. When we consider inputs from sensors like cameras things slow down considerably. Cameras deliver images to a computer via serial communications, so on each clock cycle a small fragment of an image can be transferred or analysed. So examining an image with 5000 by 4000 pixels will take on the order of 20 Million clock cycles to transfer or process an image. With a 2 GHz CPU this gives a rate processing images of around 100 Hz or lower. The point being processing sensors slows the clock rate to a much lower perception rate.

Audio sensors in computers typically operate at around 44 kHz to be able to record the frequencies humans hear effectively. Other industrial sensors can operate in the range of a few Hz up to millions of Hz. The clock signal is a bit too fast to effectively synchronise the sensory system, so typically it will operate at the slowest rate so we would end up with a synchronised sensor loop around 10s of Hz.

In humans the perception cycle we can see happening in computer games. Games are designed to run at as high frame rate as possible. At the high end this is up at 100 Hz, but lower frequencies of 60 Hz are considered acceptable. At high framerates people report that the movement appears smooth. Around 40 Hz people report jittering. Essentially this means that our perception is running fast enough to be able to detect the changes in the images at 40 Hz. At 25 Hz the jittering becomes quite noticeable for most people and below 10 Hz games become unplayable as images are not responsive enough for players to be able to control the game. Noting the performance of the perception loop, I would expect that our entire consciousness loop operates at a similar rate 40 to 100 Hz.

People often report the idea of time dragging on or being so involved in something that time passes quickly. This could be related to the change in frequency we see in the recording of human brains. At low frequency time would appear to flow faster. This typically happens in people when they are busy with an activity. This might occur as we are putting more effort into "processing data" thus slowing down the loop. Time slowing down would occur if the frequency of our consciousness loop increased. The upper rate we detect at 100 Hz.

I heard a story a while ago about optimisation in computer science: A group of programmers were attempting to make their program faster. They ran a code analysis and found that their program spent 80% of its time in a particular function. So they decided to optimise this function and made it several times faster. Then re ran their diagnostics to see how the optimisation ran. The program remained the same, 80% of the time spent was in this particular function. Turns out it was their idle loop!

This was intended as a lesson to think about what you are attempting to optimise. Amusing, but there is also a useful idea about how a program would respond. Once the idle loop was optimised they would find that the program was more responsive. With a doubling of speed in the idle loop it would be twice as fast reacting to new inputs.

We have covered here the idea of cycles in consciousness and how this relates to both artificial and biological consciousness. What we find is that the cycle frequency of the loop is related to our sensors and the amount of data being presented to the system. Some sensors such as audio also rely heavily on temporal ordering to be able to make sense of the sounds.

2.10.1 The Binding Problem

The binding problem described by David Eagleman is basically a synchronisation problem using multiple sensors. How the brain achieves this we aren't completely sure. To achieve this in a computing system is fairly easy. All we need is a "queue" which can relate the sensory information being received by time, a synchronisation queue. Each sensory reading, frame from a camera, audio clip can be time-stamped and a pointer added to the synchronisation queue. The queue can then be sorted by timestamp to make sure that each cell in the que is in time order. If the queue can accept sensory pointers from multiple sensors the sensory data does not need to be added at the same time, but after sensor data has been added from all available sensors the binding is complete and if we step through the que we get a full sensory experience that is synchronised. A similar process may also be possible for spatial data except using a three dimensional map to allows spatial navigation and reasoning.

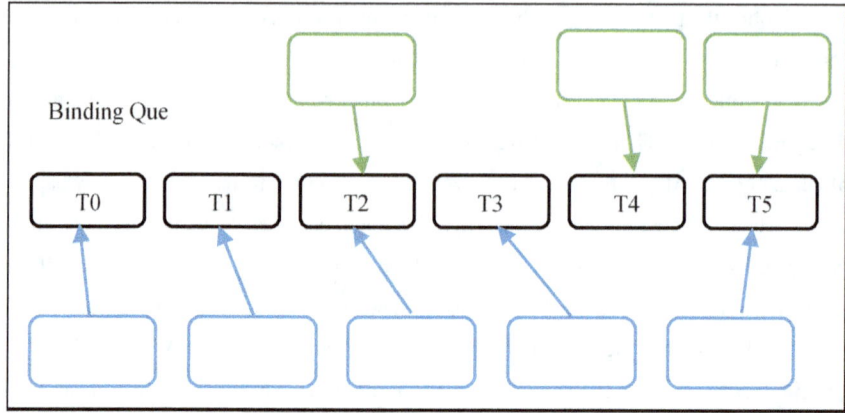

Figure 41 Schematic of a Binding Queue

In humans this management might occur in the hippocampus, which is noted for its association with memory and spatial reasoning.

2.11   Concepts

Previous researchers [Edelman, Caruthers, Baars] have suggested that we need to be able to use concepts as a basis for knowledge and understanding. I define Concepts as ideas that are not necessarily accessed via language. This allows them to be used by animals that do not necessarily use language.

From an evolutionary perspective the concepts that are most likely to emerge initially I think would be:
- Food
- Threat
- Mate

These three concepts would be the basis for most life forms to survive and reproduce. Other concepts would develop as needed and my guess is that when social creatures evolved we would start needing more concepts and eventually language.

2.12   Language

Language is a common topic in previous research that I think is well covered by other researchers [78]–[81]. There seems to be some disagreement over whether consciousness requires language, but I and many others agree that consciousness is required for language to operate. Given the existing coverage on the topic I won't go into much detail here, but I think I can add something to the conversation.

From an AI perspective the way language is analysed is to break up the sound stream into "phonemes" and store them in a "que". From there the phonemes can be grouped together to determine which words are being used. This can be fairly easily accomplished with a set of neural nets. Each network would be attached to the que offset from each other, so that each network can operate independently and come up with an interpretation of the incoming sounds. Then the networks can compete as to which interpretation actually makes the most sense.

Figure 42 shows this concept with the networks interlinked. This allows multiple phonemes to be grouped into syllables, then words then sentences. The important part is that using this type of structure each phoneme can only be used once, but multiple sentences can be generated each only using every phoneme only once. Then another part of the brain can decide the meaning of which sentence is most appropriate.

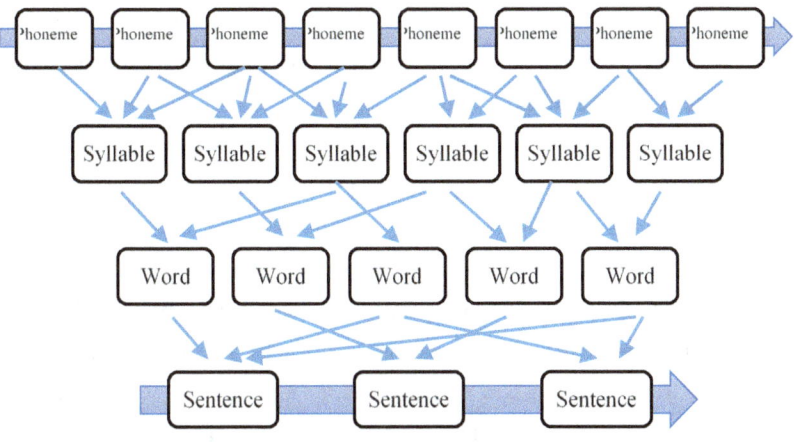

Figure 42 Language Que

As discussed in section 1.4 AI has manages to develop systems that can both analyse and synthesise language, to the point where they are competitive in Turing tests. In other words they can convince an unknowing human being that the AI is human.

2.13   Learning

Memories help us to learn new skills and thus can adapt to new situations and ultimately survive better. Learning is closely related to memory as we need to learn new skills and remember old ones. Learning represents the intersection between not having memories and having new ones. In a way we can describe learning as recoding new memories.

In the training industry four stages of competence are recognised:

1. Unconsciously incompetent
2. Consciously incompetent
3. Consciously competent
4. Unconsciously competent

If we are unconsciously incompetent we are not even aware that we don't know what we are doing.

Consciously incompetent means that we are at least aware of not be able to accomplish a task even when we are actively attempting it.

Consciously competent means that we are able to accomplish the task when we are actively thinking about the task.

Unconsciously competent is the final stage where we have refined the task to the point where we do not have to think about the task while performing it. This is often described as "muscle-memory". This means that our muscles have been conditioned to perform "the" task to the point that it requires less direct attention.

In respect to consciousness this is obviously interesting as consciousness is directly involved. Again this is a situation that allows us to investigate consciousness and how this affects our behaviour.

In terms of the design learning would be involved when "creating" a new skill or reflex action and updating an old one. For this to occur we can imagine that an old skill is "loaded" into consciousness and its performance examined. During the simulation we can actively examine the difference between how we find our skill operating and how we want our skill to operate. Presumably, as we examine the output of the simulation and the actual operation the differences can be used to update the behaviour and "learn" the new behaviour.

Baars (p362) suggests that learning alters the experience of the material learned.

The design needs to learn how to "create" or "reproduce" any stimulus learned or memorised accurately enough that it can recognise its own qualia correctly.

Programmable computing systems like FPGA's and Neural Networks require a "field" of programmable elements. FPGAs are gate arrays that allow various functions to be implemented and connected together giving a flexible way to create a dedicated piece of computing hardware. Neural networks contain a "field" of Neurons each with a set of connections to other neurons. The field in this case allows the weights between the neurons to be adjusted to create preferences for different connections under different circumstances. The common feature here is that the field can be modified, allowing the system to "learn" new information or rules.

As the human brain is more or less a neural network, I suspect that a similar process is at work. When we learn new information neural connections are updated and new patterns and rules can be recognised and reproduced.

2.14 Goals

Goals are one thing that people eventually can focus on consciously. The general procedure is that we create a goal image and use control theory (run, measure, adjust, repeat) to minimise the difference between the goal state and the current state.

I suspect that creatures that do not use goals might have limited consciousness and simply rely in instincts and reflexes to achieve their biological goals. As mentioned previously once we have the capabilities for language consciousness must certainly exist. For language to work we need to be able to understand abstract concepts. As such I suspect consciousness must be developed earlier in the evolutionary chain. Having abstract concepts for goals might be a good place to look for that point. The ability to have goals, specific targets allows us to focus our actions to achieving the goal. This can potentially reduce the energy requirements of an animal while at the same time allowing more capabilities to be developed for specific goals to be achieved.

In the design goals are not specifically supported, however as mentioned they can be thought of as abstract concepts. As there are also other processes in place in the design to allow comparing state information to recognised information these processes should be easily adapted to generate a desired goal state. Once this is operational we can also use the control system idea to see if our goals have been achieved.

The important part is that even though we do not necessarily have direct support for goals they can be indirectly supported by higher level conceptual functions. This may be similar to what Baars describes as frames.

2.15 Consciousness

The concept of consciousness is the goal of this design. The question we have is: Is it possible to come up with a process based design that would allow consciousness to occur. So based on the ideas presented in previous research, a detailed review of the processes operating in the brain and some lateral thinking this is the core concept that has been developed.

The consciousness component can be extracted out of the design as shown in Figure 43. From there we can state: Consciousness is generated from a feedback loop from sensory input being used to update an internal simulation of the self and world around it. The simulation can be used to both direct the focus of the consciousness and to make decisions about actions.

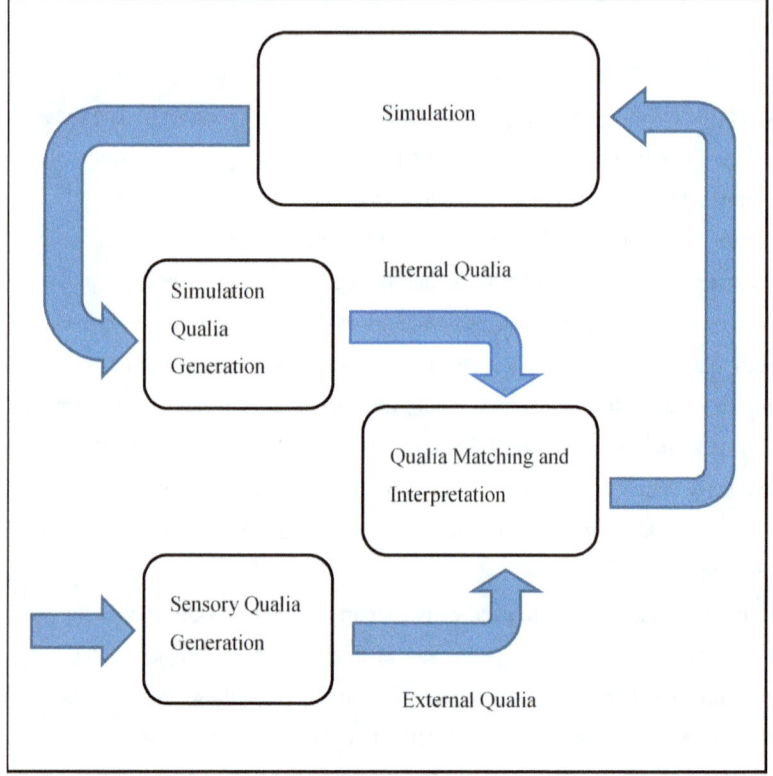

Figure 43 Detail of Consciousness Simulation

The focus of the consciousness can be controlled by generating "inverted signals" and combining them with input sensor signals. If a signal and its inverse are added together the signal becomes "zero". This means that the input signal can be ignored and something that is interesting can be amplified.

So from an evolutionary perspective this could have been generated from a predator prey situation. For example, a lifeform with some visual sensors (eyes) may be able to examine its environment. The parts of the environment that do not move can easily be filtered out by subtracting the previous image from the current sensory input. This also results in any moving objects being highlighted. This is useful for a prey lifeform as it can then look for other things in the environment that are moving, such as predators.

As I understand the operation of human eyes the subtraction process is actually done in the eye itself. We actually see in 6 different ranges: Red, green, blue, movement, edges and blob.

I expect that all of our sensors have this ability so that we can focus on or ignore sounds. The interesting thing is that if you invert sound and listen to it, the inverted signal sounds exactly the same as the original. I'm guessing that as this part of consciousness evolved it allowed qualia to be generated. In general terms focusing on sounds and ignoring others allowed us to create an internal voice. An examples of our consciousness generating qualia would be internal dialog. Essentially this allows us to hear a voice when we are thinking. A person is aware of the words, they want to say, but are only experienced by themselves.

The next idea we can look at here is when things stop working properly. Such as schizophrenia, where people literally hear voices. From looking at Figure 43 Detail of Consciousness Simulation, there are 2 places qualia can be generated:

1. From external sensory inputs
2. Internally from the simulation

From this we can create a hypothesis for schizophrenia. Some change in the processing occurs so that the brain cannot distinguish between internally and externally generated qualia so for example people who hear voices cannot distinguish the difference between "voices" that are generated by their own internal simulation and actually heard from external sensors. Notably this is similar to the theory presented by Fernyhough from Voices within.

One thing that may be noticed from the model is what isn't covered by consciousness. I have deliberately placed a few components outside of the consciousness loop including memory some areas of sensory input and reflexes/values.

Memory as previously discussed only becomes conscious when people bring a concept from long term memory into working memory. The memory would be conscious when be maintained by the perception and simulation components of the consciousness loop.

Sensory processing is again more complicated as we are obviously aware of the data we receive, however we are not aware of the instinctive processing that occurs. For example we see in three colours Red, Green and Blue from the rods and cones in our eyes, but we are also sensitive to "movement", "blobs" and "edges". These can be generated from examining adjacent "pixels". This processing could be performed on the eye itself, which we would not likely be conscious of. Sensory masking would significantly affect what we can be aware or conscious of and the way it is implemented in the design it also allows our consciousness to be focused or directed to where our awareness is required.

Our Reflexes and Values which may be incorporated in what psychologist call the "self" are also probably not directly conscious. However, we may get an idea of what our values are by simulating a set of conditions and examining our own feelings. The other part of the self would be our memory and stored structure of our physical self. My guess is that with the combination of our inherent reflexes, internal values, our self-model and associated understanding would result in our understanding of our "self" in a psychological sense.

From comparing Figure 14 and Figure 43 we can see the extents and limits of consciousness and by looking at its operation gain an understanding of how they interact.

The conclusion I come to from this design is that consciousness can be described by a fairly simple process.

That said, I am not completely sure that this idea equates to consciousness. It is, I think, very close.

So in the rest of this book I will be attempting to expand on the operation of this model and make detailed comparisons to both biology and other research to test this idea.

2.16 An overview of operation

As we have established a model of consciousness we can look at its operation. To examine the operation we start with the external stimulus and trace the effects through the system to look at what outputs result. Refer back to Figure 14 Figure 15 and or Figure 43 for this discussion.

Any external stimulus that can be detected results in sensor inputs to the relevant sensory appendage. From there any sensor data is processed and transferred on to be masked so that relevant stimulus can be examined. This allows the differences in sensory inputs to be further processed to be made conscious. Sensor processing generates qualia that can be compared to simulation qualia. The incoming sensor data is also fed into reflexes to allow any necessary reflexes to occur without consciousness.

The simulation qualia is than analysed in an attempt to recognise what is being sensed. The recognition is built up over time with iterations of the consciousness loop. The recognition templates are built up from different sensory experiences to separate out "objects" from the background which could also be classified as general world objects (the ground, sky etc.). So the recognition system needs to be able to handle objects from different points of view and also be able to relate different sensory experiences visual and auditory to the same object, which is where conceptual recognition occurs.

These recognition templates can then be "loaded" from long term memory into working memory and "operated" as understood to make predictions about where objects would move or how interactions will occur.

The results of the simulation can then be interpreted as qualia and matched to incoming sensory data. This allows the simulation predictions to be examined. This has two outcomes. First is our understanding of the world can be updated if our understanding does not match with what we see happen, this would be called "Learning". Secondly we can attempt to problem solve and **not** make instinctive reactions. This suggest, the consciousness loop allows "Conscious Decisions" to be made about actions.

One of the main features of global workspace is the global aspect of the theory. This idea suggests that sensory data or workspace data is broadcast globally around the brain. This idea has been developed from the fact that sensory data does appear to cause global transmissions throughout the brain especially when the stimulus is "novel". Novel in this case suggests that the stimulus is new to the observer. When we are attempting to identify the sensory data it makes sense that if memories are distributed and close to where they are processed that all of these areas are checked to see if the data stored is relevant to the incoming stimulus. So it makes sense that incoming stimulus is broadcast widely, however it does not make sense that stimulus be examined across sensor modes, for example visual stimulus is examined by auditory systems. It would never make much sense and waste a lot of energy to attempt to process. That said the higher level processes may still require further processing. This means that we first trigger direct sensory memory, then we attempt to identify this memory: is it part of ourselves? Is it from the external world? Further along is a symbolic analysis or object recognition: We see a moving object externally: Attempt to isolate and identify what the object is. Turns out this thing moving towards us looks like a dog. Further up we have what I call a conceptual analysis that allows connections between senses and some forms of analysis. This allows concepts to be defined by multiple sources of sensory data. This than allows to cross over between senses and examine other sense data for related stimulus. In other words: Oh I can see a dog, Oh look it is barking as well. Going back to the sensory processing change and the "global broadcast" when each sense data is identified the brain would examine potential symbols and conceptual, perhaps even abstract concepts. So as the higher level analysis occurs the stimulus would be transmitted further up the processing change. Then when the objects are identified these symbols could then be transmitted downstream into the simulators to predict the stimulus we sense.

All this comes back to the point that I don't think that this stimulus transmission is "global", any data needs to be transmitted to the areas of the brain that need to process the associated data, but not everywhere. That said we do end up with cascading signals up and down through the analysis and simulations that would relate to most of the operations of the brain.

In terms of the sensory mask much of the stimulus would be halted at this point. In other words we only process the changes in data that are not currently masked out. It has been demonstrated by Baars, Caruther and others that we may be unconscious of stimulus that occurs in our field of view especially while we are focussing on another complex task. The more focused we are on a task the more the mask is required to ignore irrelevant sensory data, especially data that does not change over time. That said using the difference system for masking means that the changes in stimulus can be transmitted. The benefit of this is that when things do change in the real world a signal is created which can get past a mask and alter the focus. This could be handy from an evolutionary perspective when our ancestors were prey for larger creatures we needed to be aware of.

On top of all this we have active or functional knowledge (understanding): we know of gravity, things fall "down" when we let them go. From this using basic physics and math we can predict an exact path an object will follow when released. Biologically we can build similar reflexes so that when someone, for example, throws a ball at us we can predict where the ball will move to and thus catch the ball. This is the basis of making predictions: our knowledge can be functional, this allows us to make predictions and can alter the outcome of expected results especially of physical phenomena. These predictions can be used extended to other areas including interpersonal relationships, social relationships, even abstract concepts like mathematics, philosophy etc.

2.16.1   Awareness lag

Researchers have noted that we can identify when people have made decisions long before we act on them. We often don't realise we have made a conscious decision until after it has been made. One part of this is that we need to simulate the situation to make a decision, which takes time. The other part may be that even once we find a final solution we still take the time to examine alternatives before finally acting on them. Either way a significant amount of time can occur between input stimulus and processing the simulation before people finally act on a decision.

2.17 Stream of Consciousness

The basic idea is that we have a sensory system that is continuously receiving information and attempting to analyse and identify what it matches to in its "database". The identified concepts are generated in our working memory, simulated and matched back to the original sensor data and analysis. Also, even when our sensory systems shut off images and sound is generated by our brains that continues to examine concepts we know about, our memories.

So if we allow the model to examine closely related concepts we end up having a wandering "Stream of consciousness". In the technical sense every concept has related concepts. In software systems every "class" has associated methods, properties and exists in a hierarchy of classes. For example we could be looking at a ball, which we can think of as a type of "sphere", this would be an example of a class hierarchy relationship. The ball would have many attributes. For example might be coloured red. It will have a physical size, a mass, an elasticity. The methods would include how we use the ball, bounce it or throw it etc. and more complex concepts like play games.

From these ideas our consciousness, the term I use is, "explores" the connections between concepts. What else is red? What other games can we play? Each time we think of a related concepts we start to become aware of other things that are again related. This leads to our minds wandering through a network of concepts examining and updating links between concepts. This could be done in any sensory mode, but the popular ones are visual and aural (sound i.e. we talk to ourselves). As far As I can tell these ideas are fairly similar to those described by other researchers especially Dr Fernyhough and his "Voices within".

2.18 Comparing to Other research

In section 1.3 we reviewed a few samples of current research on consciousness and cognitive science. In this section I will give a brief overview of what I got out of their various works and focus on a few I think might be helpful.

From Dennet's work the idea of emergence seems to come into play from his concept of competence without comprehension. While I do not agree with his conclusion that we are basically zombies, the idea of emergence seems like a good starting point for understanding intelligence and how it could produce consciousness.

Dennet and Pinker both examine evolutionary processes that have formed our minds. So evolution seems to be a common point of agreement. Whatever process causes consciousness it must be able to evolve from a simplistic system to a much more complicated system.

While Searl does not produce a model of his own the questions he generates are relevant to understanding consciousness.

While Chalmers takes a roundabout philosophical view he does come up with some questions to ask and some properties we should expect from consciousness.

My feeling is that much of the other research shown here has much in common with the design I have developed. The following sections go into more detail with the prevailing theories of consciousness and "the design". The processes described in the design should be common to both artificial and biological consciousness. As such, it should be possible to compare the expected output and operation of the design to biological consciousness and any features of artificial consciousness.

2.18.1   Panpsychism

According to Panpsychism everything is at least potentially conscious.

{AAAArgh my Chair is trying to eat me!}

Some of these ideas have been developed from quantum physics concepts of entanglement, decoherence, and a few others. This has been considered by a few notable physicists and popularised among many others. From the design developed in this chapter we can safely rule this out. For a piece of matter to be conscious it needs to process information. It needs to both analyse information and generate information and we can see that basic matter does not perform any of these tasks.

{Up-side: My Chair is not trying to eat me. Sadly, it also means we probably will never develop cool Jedi Powers. Sigh...}

There may be more complex processes going on in consciousness that require quantum effects or panpsychism in some way, but we can say that these processes are not necessary for this design to operate.

2.18.2   Koch "The Feeling of Life Itself"

In chapter 13 Koch examines the CPU and demonstrates how to examine the integrated information of a CPU is zero. To be specific the part of the CPU examined is called a logic unit. These part of the CPU do not store or retrieve information, they simply calculate output numbers from given inputs. In this case I agree the Logic unit will not have any feedback loops that store information. However, the logic unit is only one part of the CPU, there are also memory systems and command sequence control. These again may not have integrated information, but in conjunction with software feedback loops can be created in memory and control loops. By the IIT standards, computer software could have a non-zero Phi. Using Dennet's idea of competence from distributed intelligence we can also get to the idea that components do not necessarily have to be conscious for a group of cooperative agents to create consciousness.

Koch presents in his Integrated Information Theory (IIT) of consciousness five postulates that describe the operation of consciousness.

1. Intrinsic existence
2. Composition, has Structure
3. Information
4. Integration
5. Exclusion

I find that these principles might be useful, but do not necessarily add up to a working model of consciousness. That said they are useful at examining models to see if they have similar properties. As such, we can examine the design to see how compatible both theories are together.

1. Intrinsic existence

The Simulation – Qualia loops can maintain a coherent conceptual structure by simulating outputs that can then be interpreted by higher level processing (Symbolic and Conceptual) and reproducing these signals in the simulation. This loop can create and maintain information, hence the loop itself could be considered as intrinsically storing information. Koch examines computer hardware and demonstrates that it does not have intrinsic existence. This can also be said of a single neuron. However, again in software the processing that occurs requires 2 or 3 components. 1 the qualia comparator, 2 the simulation, and 3 the higher level symbolic/conceptual recognition system. The symbolic recognition system could be considered part of the simulation. That said it is not possible to remove any of these components without losing the loop and the information that progresses around the loop.

2. Composition, has Structure

The information contained in the simulation and its outputs contain the structure of concepts or ideas in memory. The structure comes from the various stages of processing. First sensory information of the self, secondly sensory information about the environment, next is symbolic information such as words or images of objects, then conceptual and potentially abstract concepts. This structure also includes the individual concepts and their related sensory information.

As a side-track: these concepts are typically based or at least accessed via sensory information, which is also consistent with Carruthers discussion in "The Centered mind".

3. Information,

The simulation and its outputs contain information about whatever concepts are being simulated in working memory. The description of information is that it is "A difference that makes a difference" in the system. The simulation –qualia loop maintains information. If this loop stops cycling the information the design loses consciousness of said information. As such, we can think of the information cycling through the loop as a difference that can be used as a basis for making decisions about what actions to make.

4. Integration

The structure of the design cannot remove any component and still maintain the simulation loop. This means that the communication between the components is essential to maintain the sensory loop.

5. Exclusion

The simulation has many limitations on what can be part of it. It is all certainly collected inside a single structure and is not omniscient. In fact, with capabilities similar to a human, it would probably only be able to maintain a few concepts, our 7 +/- 2 items in working memory.

I don't know for sure, but I think it might be getting close.

In "The Feeling of life itself" Koch describes his use of a sensory deprivation chamber. For humans this can be a relaxing and somewhat spiritual experience that many enjoy. It seems to be that the lack of external stimulus allows a person to almost lose the entire perception of "self".

In comparison "the design" would be able to generate a self-sustaining set of qualia even when all sensory information was blocked. It would basically generate qualia for "I am sensing nothing at present" for the self-simulation. It would not necessarily stop any internally generated inner voice, but one had run out of things to discuss with oneself it (or they) might shut up and let you experience the gentle quiet of the void. The consciousness would not stop existing, but would remain in a dormant state until external stimulus or some other motivation spontaneously occurred.

Does this qualify with Koch's experience in a sensory deprivation chamber along with his other criteria for integrated information theory? Again I don't know, but I think it is getting close.

2.18.3 Peter Caruthers "The Centered Mind"

Carruthers established that the majority of our mind uses a sensory basis to operate. This agrees with the design in principle as the design primarily considers how information is transferred around and analysed. As such, the basis of most of the information moving around the mind would have a sensory basis. Even conceptual or abstract level processes require an audio or visual cure, a spoken work or the visual equivalent, to access the concept in the first place. This idea is compatible with "the design", as it follows the sensory perceptions as they are processed in the brain.

It also makes sense that there is a primarily sensory basis of the brain as even abstract concepts have a non-abstract visual or sound (word) associated with it. The only way for people to be able to use abstract concepts is to remember some kind of access point i.e. the word that represents the concept.

So, short version I agree with many of the ideas Caruthers presents and I don't think the design as presented contradicts his research.

2.18.4 Philosophical Zombies and Autonoetics

Chalmers wrote of philosophising about zombies. Beings that were not conscious and yet could operate exactly the same as a conscious human being. My feeling was we needed a better definition of what consciousness is to be able to distinguish what Chalmers calls a zombie.

We can consider that a creature operates based purely on reflexes with no consciousness as shown in Figure 27 Reflexes. When a reflex agent such as this receive a stimulus, they would output a related action immediately with no thought, similar to Figure 28 Reflexes to actions detail. So the question becomes: Can our reflexes become complicated enough to take reproduce conscious behaviour? What I have come up with is: Yes. Especially, in a computer system.

Making a prediction, it seems to me, almost the same as "Autonoetic", the ability to put oneself forwards or back in time. We as human beings can put ourselves in a simulation of past or future event and make a prediction of how we would act. I don't think we do this in biology as we can't do reflex calculations of the complexity required to simulate ourselves. This is likely due to the energy requirements to make these calculations and the fact that computation takes time, something creatures in biology don't have, especially when faced with a threatening situation.

A computer, however, can make complex calculations by "reflex". As such, they do not require the internal simulation and can still create realistic behaviours. This means a computer can act as a human being (theoretically), but still not be conscious.

This allows us to distinguish between a zombie and a conscious being from the following:

A reflex agent acts purely on instinctive reflexes to react to the world around it, whereas a conscious living being uses an internal simulation of itself and the world around it to make predictions about actions to be able to make decisions.

We can think this through a bit further and show that reflex zombies will "not" have the same behaviour as a conscious entity, unless it was a robotic or computer system. Given that Learning is accomplished with consciousness I would expect Zombies to ignore or avoid novel situations.

A biological entity has a cost to thought. Both time required to calculate actions and the energy for such calculations means that a biological entity would likely attempt to minimise "conscious" thought. So we are only going to think heavily about things if there is a significant benefit to the thought i.e. they don't die. A computer system has no inbuilt limitation for energy use so the algorithms used can be energy and time inefficient thus allowing them to be far more accurate.

2.18.5   *Voices in Our Heads*

Before I read Fernyhough's work, I predicted that we would have a region of the brain that produces one or more voices that is also connected to the speech processing part of our brains that produces speech. This gives us a feedback loop between our internal voice to our speech processing, allowing us to decode our own voice which would also be reinforced by hearing external voices. If you think about Koch's IIT theory how essentially feedback loops can be used to store information in consciousness, this kind of process is shown in both the design and, from Fernyhough's work (if I understand it correctly), the interaction between Wernicke's area and Broca's area in the brain.

The design also allows us to consider how this might break and result in the same problems we find with inner voices not being recognised or even full blown split personalities. These topics will be discussed further in Chapter 4.

2.18.6   *David Eagleman's Brain*

Eagleman does not generate a model of consciousness, but he does describe the operation of the brain in similar terms. "The brain creates a simulation of the world around us", "The brain is used to make predictions about what will happen in various circumstances". Again this is another source of research that shows many similarities to the design and gives me more confidence that the design is a sensible idea.

2.18.7 GWT

Comparing to "The Design" which uses multiple feedback loops through a simulation several similarities can be seen. "The Design" uses a simulation which requires the use of memory structures.

The feature of "focus" Baars relates to memory. "The Design" uses a mask and subtract idea as part of the consciousness feedback loop that would also allow sensory input to be filtered and as such focused on. From what I understand of GW "The Design" can be described as an extension of the global workspace idea. The "simulation" can be thought of as a more specific version of the workspace. The simulation must be in working memory as GWT suggests, but it is also active in that it creates predictions about what might happen in a given circumstance. Typically the simulation will be the current "state" of the conscious entity, but it could be used to imagine potential actions, or even just to daydream. The idea is that this provides two purposes. The first is at a low level to predict the incoming sensory data and thereby reinforce the incoming data. The second is to be able to predict high level situations such as: Is this thing moving in front of me going to eat me?

The problems I have with GWT are the idea that information is transmitted globally throughout the brain. From examining the processes involved, incoming sensory data may cause cascading analysis throughout the different processes of the brain from pure sensory analysis up to concept and even Abstract analysis to determine what any sensory input can be recognised as. These transmissions are broad but not exactly global.

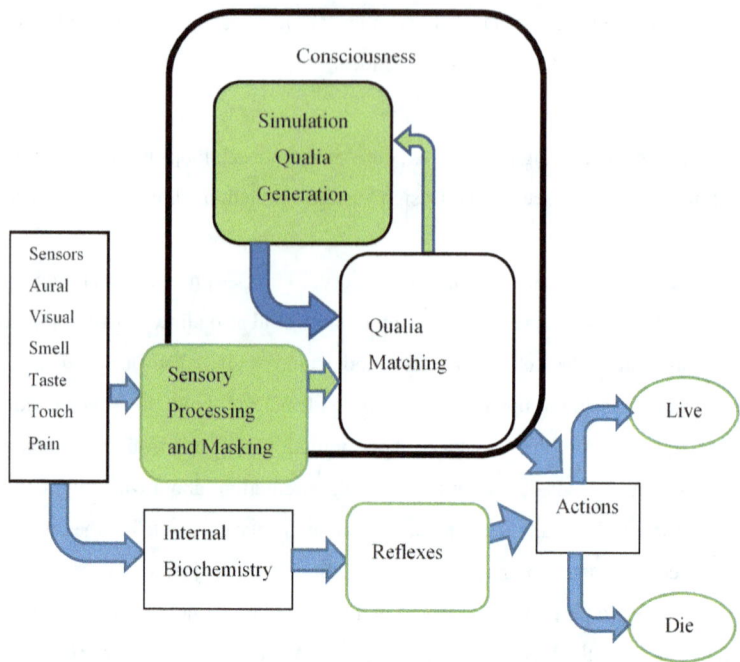

Figure 44 Design of Consciousness with GWT Highlight

Frames in GWT are used as a basis for managing tasks, however I am not sure if this requires special hardware for the brain to process or if it is simply a part of the inherent nature of information that we normally deal with. It seems to me that frames are more describing the contents of consciousness as opposed to the nature of consciousness itself. It is certainly a part of Psychology so whatever form consciousness takes it needs to be supported at least in humans.

2.18.8   Stanislas Dehaene "Consciousness and the brain"

My feeling is that much of Dehane's experiments and results follow the design presented in this book. As mentioned in the previous section I feel that the Global workspace model is a start, but does not cover the full operation of the brain and more specifically does not show how information is processed to determine actions from sensory information. That said Dehane presents some very exciting research much of which is consistent with the model presented in this book. These concepts include:

- Presents the example where a person's brain was stimulated and the patient recalled the face of Bill Clinton. This shows that a set of neurons are trained to recognise a specific person. My feeling is this represents long term memory as hypothesised in section 4.1.8.
- Close networks of neurons especially in the occipital Lobe, where visual processing occurs. Vs image processing. Typical image filters in computing require calculating the difference between neighbouring pixels, as such I would expect a neural network approach to image matching would produce a similar neural structure to what Dehaene describes.
- The observation of "ignition" a cascade of activation throughout the brain referred to as the P300 wave when people successfully become "conscious" of stimulus. The P300 wave is described as an inhibitory signal. My feeling is that this might be a "sensor mask" as described in section 2.2.1. This mask would be filtering out as much information as possible to allow the brain to focus on the most relevant information to the situation. My guess is that the areas not masked would be highlighted by short term memory, everything else gets filtered out so that the person can focus on the interactions between the unmasked ideas.
- Dehaene gives a detailed description of Schizophrenia noting that the performance of neurons linking distant areas of the brain and hypothesises that schizophrenia as well as some drugs interfere with the operation of these neurons. He suggests that this reduces the ability of people to make corrections between their internal model of the world and the incoming sensory information, which leads to a feeling that "something remains to be explained" which would "trigger an endless avalanche of interpretations". The design is used in section 4.4.1 to model this description.

Overall Dehaene presents a great deal of evidence from medical research that I find consistent with the Design.

2.18.9   HPC

HPC has been successfully applied to visual perception of two dimensional (2D) images. These applications typically use templates or models of objects to perform recognition tasks.

One of the limitations in the research that I can see is that humans have two eyes which results in stereo vision and depth perception. My first job out of university was with a company that developed photogrammetry software that used stereo vision to generate 3D models. 3D Models can be used to examine the relationship between objects in 3d dimensions, specifically this can be used to determine occluded objects. I suspect that the integration of stereo vision into a HPC model may help explain some of the difficulties identified in HPC research.

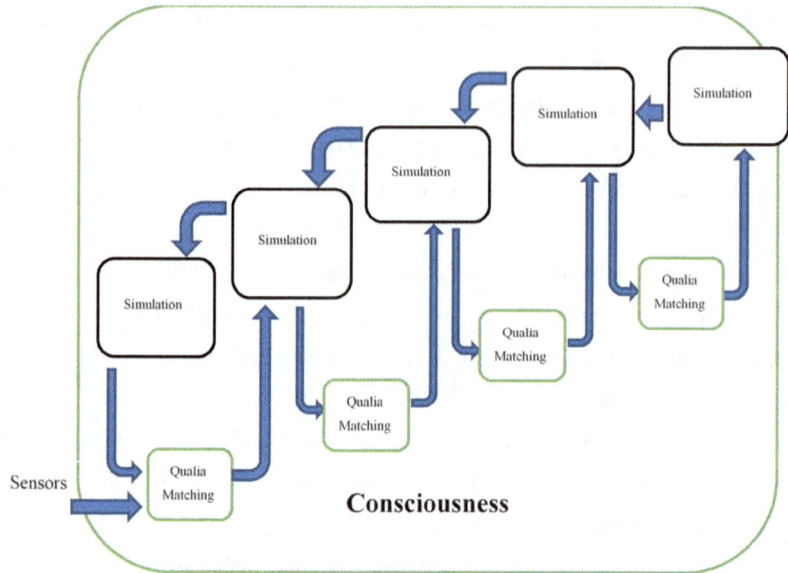

Figure 45 The Design with HPC similarity

The design I propose here is similar in many respects to HPC in operation, I use specific processes for the hierarchy, from sensory, Self, World, symbolic, conceptual and abstract. Also HPC uses a comparator operation whereas I define a simulation along with the comparator. The good news here is that both theories describe a similar process where the concepts identified in the higher levels of processing are transmitted down through the simulation/ comparison system as shown in Figure 45. As such, I am satisfied that both theories are consistent.

2.18.10   Bayesian Theory of the Mind

Some descriptions of Bayesian modelling show it correcting for the errors between "model based predictions" and incoming sensory information [82]. This process is described as minimising free "energy" in the system. This is similar to the idea used in The Design where incoming sensory information is compared to outgoing simulations. The differences are highlighted and used to make corrections to the simulation.

Much of the trouble I have with probabilistic models, specifically Bayesian Theory of Mind (BToM), is that it uses a probabilistic mechanism to accomplish the same task as other methods. This criticism has been noted by a few people [83], not just me. There is definite evidence that the technique produces very similar results, but BTOM does not explain the internal structure of the mind, it reproduces the results of the mind effectively leaving the mind as a black box.

This is not to say that BToM does not work or is not bringing useful insights. It may even be possible to modify or implement a BToM version of the design that operates as a human would. Also some processes in the real world such as how we estimate others beliefs in a social context may be performed by an internal Bayesian system. As we know many potential beliefs ourselves we may be able to estimate the probability other people share such beliefs. In contrast, "The Design" shows a causal processing structure that takes sensory input analyses it and makes decisions and performs actions based on the analysis. For me a satisfying explanation of the mind includes a description of how what processes occur to generate consciousness.

2.18.11 Choices

After looking at the conscale my understanding of consciousness changed. The conscale as it is defined derives from in many ways emotions and the types of external interactions we have with our environment and our companions. This is useful in many aspects, however I suspect it misses the mark. I suspect consciousness derives from choices in actions. At lower levels of consciousness the lifeform does not have a choice, they simply reacts to its environment.

Small creatures, such as a mouse, looking at environment has three main motivations, finding food, finding a mate and avoiding predators. So when looking at an environment it has to make a choice, go in and find food or a mate, or avoid the area because a predator is lurking there.

Once there is a choice of possible behaviours this may cause consciousness to emerge in some form to attempt to resolve when to use particular actions or strategies. Or consciousness itself may allow these choices to be made. Either way the availability of choices and consciousness seem to be related.

The point is there is a choice in this environment, there must be a mechanism for the choice to be made in the mouse. It can be argued that instincts or reflexes can determine the choice. Short version Yes, they can. There is also the condition of competing instincts. Go and save my mate from the predator or run away and find a new mate. What choices do we make when there are multiple competing options? Notably, it is also possible for more complex creatures to decide to act against their instinct, such as humans avoiding having children until they can afford them.

In general the rule I come to is: the more complicated the choices the more complicated the mechanism required to make the choice. Hence, with very similar options available with very little instinctive motivations, requires some precise analysis to examine possible outcomes. This is where the idea of the "simulation" in the design comes from. It allows us to examine the potential out comes of very similar choices or actions that have potentially significant differences in outcomes. Essentially we must make a "conscious" decision.

In deterministic terms it is very difficult to predict the outcome of a simulation without running the simulation itself. It might not be impossible, but it is extremely difficult. So if we can assume that the simulation is an emergent phenomena it allows the choice we make based on the simulation to be non-deterministic. This potentially gives rise to free will.

2.18.12 Language

The development of Language has been a major focus of many researchers on consciousness. In terms of choice, there are a lot of choices in using language. Which word do you use next? Socially how are other people going to understand and react to what I say? How to choose the different words we use?

I can imagine a situations where language was developing in the stone-age or earlier. Imagine a few people who have collected some seeds and stored them in separate clay containers, let's say a big one and a little one. One of the types of seeds is poisonous (the little one) and in the big one we have the ones we use to plant and grow crops from. You can imagine what happens if the wrong pot is used. The wrong crops grow and everyone in that tribe gets poisoned and dies. This creates the necessity of distinguishing between the seeds and potentially the pots. They would need a phrase like: "Use big pot idiot!"

So let's unpack what needs to happen there. We have at least two people who are required to communicate. They need to be able to recognise each other and their intentions (ConScale Empathic or Social). The person giving instructions needs to come up with a concept, a word, that differentiates the two options "Big" and "Small". The other person needs to be able to recognise the sounds made for the word, associate the word with the sounds and then associate the word with the correct pot.

So the person hears the word (let's say "Big"). This word needs to be recognised from memory or possibly the first person explaining what big is. This is accessed through the symbolic database and moved into the simulation (Baars theatre of the mind). We then get the second person imagining taking the big pot out to the first person and them being happy or the wrong pot and the second being annoyed. This is a fairly complicated social simulation that hopefully demonstrates how the simulation is useful. The demonstration of consciousness comes from being aware of our own thoughts and being able to choose which pot based on limited information i.e. the word "Big" or "Small". We need to consciously model the results of the sequence.

In evolutionary terms the societies that did not come up with the differentiations for poisonous and not poisonous would probably die out. This means we would end up with societies that can distinguish between poisonous and not poisonous in language and in some respects consciousness goes along for the ride. As such, it seems that language is a major influence for the development of consciousness, but I doubt its source.

Many animals can be taught to speak either directly as in birds or with some effort dogs and cats. Christina Hunger says she is teaching her dog Stella to communicate with the same method she uses to teach children to speak and learn words. The dog Stella has been trained to press push buttons that play and audio file of a word: water, food, beach, play, outside etc. Stella is capable of selecting a button to tell "her" (not sure) owners what she wants to do. Typically it is to go outside and play or beach. ON the link provided Stella is attempting to press the button for beach, but the button stops working. *{And it's adorable! Puuuuuuppppiiiiiiess!}*

https://www.hungerforwords.com

https://www.youtube.com/watch?v=FCVASZngVLc

She then goes on to press the button "Mad". To think about what Stella is doing conceptually we can get to the following thought sequence.

- I want to go to the beach. → press beach button
- Button is not working
- Check again.
- I'm getting mad.
- Find mad button
- Press mad button

The fact that Stella is capable of performing these tasks, suggests that dogs more generally are capable of relating these kind of concepts.

Also Billie the cat from the YouTube channel "BilliSpeaks" has received similar training. *{Kiiiiiiittttiieeeeeeee!!!!}*

https://www.youtube.com/channel/UCGMTesZlKa0Lokb7ZNqOJXQ

Despite the fact that most dogs and cats don't use language, these tests suggest that they are still conscious as they have awareness of themselves and are able to express this to a human.

There is also the phenomenon of Aphasia, discussed briefly by Koch [17] and Fernyhough [22]. Aphasia is defined as an impairment of language, affecting the production or comprehension of speech and the ability to read or write. Aphasia is often due to injury to the brain-most commonly from a stroke, but can also be deliberately caused by anesthetising parts of the brain, Koch gives an example where the parts of the Left cerebral hemispheres were anesthetised. The people Koch describe suffered temporary Aphasia, but he noted that they were able to describe the phenomena after they had recovered their capabilities with language. This suggests that our ability to use language does not cause consciousness directly.

2.18.13   Belief

There are a wide variety of beliefs held by people everywhere. Some more coherent based on real world experiences and understanding and some more based on our "desires" or motivations. It is a real phenomenon that people have beliefs, but just because someone believes something does not make it real.

So part of our understanding of consciousness must allow this kind of error to occur, at least with humans.

Again I see this as a choice: we decide to believe in ghosts, god, big-foot, aliens, or that the Hulk is better than Wolverine. Again just because you believe it doesn't make it true. These kind of things may simply represent what we want to be true, because they represent things that make the world more interesting for us.

For these kind of things to occur we have to be able to make a choice even when evidence or direct experienced does not support our beliefs. I suspect this is an emotional decision, because we want our beliefs to be true. This means that there must be "a disconnect" between the physical or phenomenal understanding we build up and the symbolic understanding we have of the world.

This is one reason I have separated the symbolic, physical and self-databases. It is fairly simple to discover inconsistencies internally in the same database by examining different entries of the database. Examining between databases can be much more complicated as they "store" different types of information. The different structures may be difficult to compare. Comparing two images, such as images from both eyes, is fairly simple. A mathematical function describes this comparison: Correlation. We don't even need to know what the subject of the images are. Comparing an image to a world level object is more complicated, but not impossible. We can break the image down using edge detection, blob detection, 3d projections and or template matching to determine the type of object we see. This is all called object recognition. Consider how to compare a visual image to the idea of something abstract like "honour". There are a lot of steps to break down an image and decide how much honour is represented in an image. That said many people can look at images of soldiers and define them as representing honour.

The different levels of databases may represent enough of a disconnect to allow our beliefs to over-ride what we see in the world. This allows us to make choices that may be related to a deeper understanding of the world at times, or alternatively create new more satisfying or secure illusions that allow us to face the next trauma we are exposed to.

2.18.14 *Mysterious Phenomenal Functional Reductionism*

One of the discussions around consciousness are the ideas of Phenomenology vs Reductionism and Functionalism vs Mysterianism.

Phenomenology is defined as:

"The study of structures of consciousness as experienced from the first-person point of view" [84]

Reductionism in the context of consciousness:

"Strong reductionism holds that consciousness exists, but contends that it is reducible to tractable functional, non-intrinsic properties."

"Weak reductionism, holds that consciousness is a simple or basic phenomenon, one that cannot be informatively broken down into simpler nonconscious" [85]

Functionalism:

"Functionalism in the philosophy of mind is the doctrine that what makes something a mental state of a particular type does not depend on its internal constitution, but rather on the way it functions, or the role it plays, in the system of which it is a part." [86]

Mysterianism basically says that we do not and cannot know how consciousness is generated suggesting that there is a missing part of the puzzle that will never be found.

Clearly most of the work on "The Design" is a reductionist and functional description of the components of the brain. The assertion is that we can potentially determine where and how consciousness develops in the brain. My feeling is that we can give some insight into phenomenology as well by examining what sensory information is received and analysed along with the knowledge and analysis capabilities.

The sensory information gives the breadth of the possible experience a creature can experience. For example with bats and sonar we know that they can determine distance from the delay between sending and receiving the sonar clicks, it may also be able to distinguish some characteristics of the target object, specifically the material of the target object. A hard smooth object will be able to reflect more of the higher frequency components of the sonar click and a softer rougher surface will only reflect the longer wavelengths. From this we can estimate that bats would be able to sense in some way the material of the object. Visual information we know has information about colour and texture of an object. So from the variety and nature of the sensory information we can generate an estimate of the potential capabilities of a sensory stream.

The analysis of sensory information then gives another layer of phenomenology that we can estimate. We know that animals can hear sounds but not necessarily understand the meaning of words. We can assume that they do not have the architecture to relate the sounds to the conceptual meaning. We can think of this in some way as not knowing the right language to decode the sounds. The same with visual objects, if you don't know what an object looks like you may not be able to see it in an unstructured environment.

From these descriptions my feeling is that we can build up a model of what it would be like to "be" another creature. This suggests that functionalism and reductionism can be used to generate a phenomenological description and potentially or hopefully reduce the insistence on mysterianism.

2.19   Summing up

Current theories on consciousness have examined many frameworks for examining consciousness. While this has provided valuable insights into how the brain works I am not particularly satisfied that we understand what consciousness is or how it operates. From examining the previous literature on consciousness most of them seem to miss the mark about how consciousness is generated from the perspective of data processing.

One aspect that is popular is the idea that we have an internal model, simulation or set of beliefs. These are essentially the same thing. The idea that consciousness is something generated by people is not new, in fact as far back as Von Helmholtz in 1867 [87] suggested that people generate their consciousness. The next stage that we examine the difference between an internal model of the world and input sensory information is also common to fields of research including Hierarchical Predictive Coding (HPC) and Bayesian Theory of the Mind (BTOM), and even individual researchers such as Koch, Caruthers and many proponents of Global Workspace Theory (GWT).

My feeling is that consciousness is a mechanism that has developed allowing us as humans to make decisions about very subtle choices that could potentially have a significant impact on how our lives will progress. The reflexive, instinctive and emotional intelligence of our ancestors may not have been enough for us to make the choices we needed to survive. As such, when our ancestors were faced with choices, those that developed the ability to "consciously" examine their behaviour, choose less direct instinctive behaviours and make more complex choices allowed them to survive.

I am attempting to learn how a consciousness could be developed essentially by building one. The design I have developed could be used as the basis for a computer program that could be conscious. In doing so it gives a better base structure to relate new measurements to.

The Design of Consciousness

3 Evolution of consciousness

We know consciousness exists, so how did it come about.

We know that consciousness is required before a structured or systematic intelligence can be formed. An example of a structured intelligence is language. Symbols either written or spoken have particular meanings. These symbols are created intentionally. The meaning may change over time, but the symbols must be created with a purpose in mind. This means that if a lifeform presents evidence of learning or creating systematic intelligence it probably is conscious. So consciousness must predate structured intelligence.

Where would it come from?

The conscale considers consciousness of "dead" matter with no consciousness up through instinctive and emotional life forms to humans and potentially beyond. This also follows a path that should be relatively close to the evolution of the mind.

3.1 ConsScale Levels of Consciousness

Measuring the evolution of consciousness

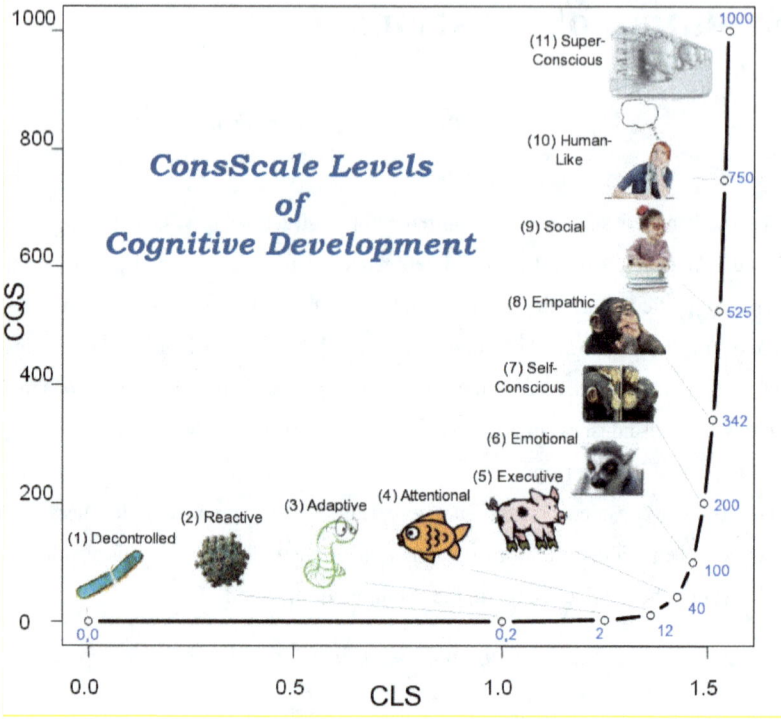

Figure 46 ConScale: Levels of Consciousness

https://conscious-robots.com/consscale/ Creative Commons Licence.

Raúl Arrabales [88], [89] developed a scale of consciousness that explains how much conscious we think lifeforms might have. This idea in many ways follows the evolutionary path that consciousness has taken.

The "conscale" levels of cognitive development describes levels of consciousness from Disembodied up to super-conscious. These levels of consciousness are described by the physical characteristics that a tree present in various life forms or conscious entities. I will give a brief overview of these levels along with a discussion about how they relate to biology, evolution and the design of conciseness. These levels are as follows: Disembodied, Isolated, Decontrolled, Reactive, Adaptive, Attentional, Executive, Emotional, Self-conscious, Empathic, Social, Human-like, and Super-conscious.

The Conscale has been developed to examine the different levels of consciousness. So at this point we are examining the differences between the design and how it could be modified to demonstrate each level. I do not totally agree with how the conscale develops, but this gives us a reference to look at how various features would develop.

It seems to me that emotional capabilities are a central theme of the conscale. It seems to me that this is a biological approach despite the fact that the conscale is described in terms of computational entities such as agents and set theory.

The design notably does not focus on emotion. From the functional perspective emotions would be a reflex interpretation of sensory inputs to create reflex actions, "fight or flight". In this context the "emotional" nature of the conscale is more relevant to biology and not consciousness directly.

In summary "the design" shows that the emergence of consciousness does not require an emotional basis. That said it is probably true that emotions emerged or evolved in biological entities at the same time as consciousness. Despite the differences in approach the conscale can still be used to examine how "the design" could be adapted to each level of consciousness.

3.1.1 *Disembodied*

All life as far as we know has developed from a presently unknown process to form life. As far as we know basic matter itself.

[Actions]

Figure 47 Disembodied

The Disembodied level of consciousness represents basic matter which only has direct connection between actions and its environment. It does not think, reactions are purely physical and is not conscious.

What does it feel like?

Absolutely nothing. No sound no pictures nothing, not even the absence of sound or vision

3.1.2 *Isolated*

Isolated refers to the basic biology of RNA or DNA. These again are not complicated enough to be able to have the capabilities required for consciousness, but they are the basis of life that we know of.

Figure 48 Isolated

Isolated entities are basically just chemistry. Any actions would be purely based on chemical interactions with the rest of the world.

What does it feel like?

Absolutely nothing. No sound no pictures nothing, not even the absence of sound or vision Everything is still blind chemistry at this point.

3.1.3 Decontrolled

Decontrolled is the first level that the science community considers a living organism. Basic single celled creatures that simply exist but live by absorbing chemicals and energy from their surroundings. They are defined as having sensor and motor controls but nor relationship between these characteristics.

Figure 49 Decontrolled

Decontrolled entities would be life forms that have limited sensors and "food" processing internal biochemistry to generate energy, but no internal link to their actions.

What does it feel like?

Absolutely nothing. No sound no pictures nothing, not even the absence of sound or vision.

But at this point we have movement. A creature can move although probably not with any particular goal or destination. This would enable a creature to move where nutrients are available but not towards or away from anything in particular. The sensors at the most basic levels are still chemical sensors.

3.1.4 Reactive

Reactive organisms again have sensor and motor controls, but at this level do have connections between the sensors and motor controls. This allows them to survive by moving towards food or away from some threats. They do not have complex nervous systems for memory. Viruses or simple amoeba could be described as reactive organisms.

Evolution of consciousness

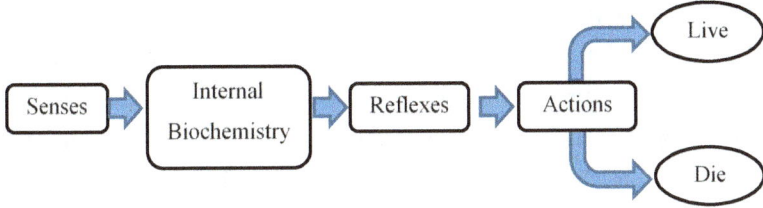

Figure 50 Reactive

Reactive life forms finally have a connection between their limited external senses and their output actions.

This is where reflexes to move would be developed. My guess is that this is when we would start a predator-prey arms race. When predators, at this stage the most basic single celled creatures begin to start eating each other. Prey that could move away from predators would be able to escape giving an evolutionary advantage. Predators would then need to adapt to catch prey, thus starting the arms race.

What does it feel like?

Finally we have some action. Once we have a connection between sensors and actions we can start talking about experience. The defining feature would be the development of a nervous system. Creatures as simple as worms have a nervous system. If you cut them it has reactions of pain. These reactions are useful if you are attempting to get away from something that is attempting to eat you. Also with a nervous system you can start feeling the external environment, so feeling of hot, cold, dry, wet, and possibly pressure would be possible. With a more complex nervous system we get internal feeling of hungry, tired.

These senses would allow us to have a continuous immediate experience of our own internal senses, but without a map between the sensors they would most likely to be general non-local feelings.

Reactive lifeforms could cover a large range of senses. And the more complex the senses the more we have a need for a self-model. The self-model would allow the different senses to be connected and coordinated together. As such, we could expect creatures to have immediate sensory experiences, internal physical feelings, and eventually an ability to coordinate between senses. Without a significant memory or adaption it would not be able to "recognise" previously encountered creatures or situations so they would repeat previous behaviours that proved ineffective.

3.1.5 *Adaptive*

Adaptive organisms can be considered "adaptive agents" in computational jargon. They still require the same sensorimotor control systems as previous levels, but additionally they have the ability to learn new behaviours. This makes them useful for simulating behaviours, but not enough capabilities for consciousness that we recognise.

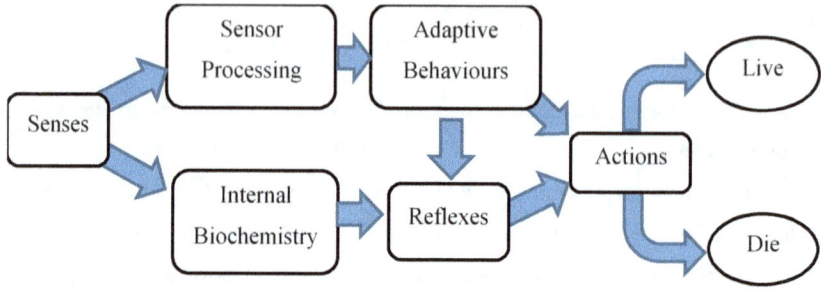

Figure 51 Adaptive

Adaptive life forms must have reflexes and must be able to alter their behaviours or at least provide alternative behaviours. This could be accomplished by providing a separate path for the sensor data. If the sensor data is processed and mapped onto a model of the creature, it could be a basis for a self-model. With the self-model they should have some form of consciousness, but my guess is that they will only "live in the moment", so at this point we can start considering these creatures as anoetic.

Once we have the arms race evolutionary pressure to adapt would generate two new features: sensors and a way to combine the sensory information to make decision or behaviours that allow the lifeform to adapt to its surrounding.

What does it feel like?

Once we have a memory we can start recognising other creatures and situations. We would expect that memories would develop of predators, food and potential mates and where these things exist.

Part of the reason I consider emotions part of reflexes is that as soon as memories exist I would expect associated emotions to develop. Specifically, fear of predators and desire for food and perhaps lust for mates.

3.1.6 *Attentional*

Attentional agents have new features of memory and focus allowing them to develop specific behaviours such as "attack" and "hide". These types of agents may also develop basic emotions such as fear or desire that would influence their behaviours.

Executive agents are defined by their ability to manage multiple tasks and "set shifting". The emotional capabilities of living organisms at this level of consciousness seem to drive their actions.

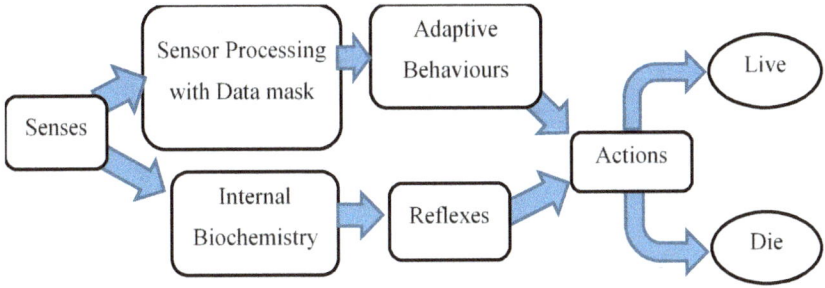

Figure 52 Attentional

Attentional lifeforms require the concept of the sensor mask. This allows the sensors to be used and some of the data to be ignored, which would allow the creature to focus on specific sensory information.

Once we have a few sensors we might need to direct our attention to the most relevant stimulus. This allows creatures to make decisions based on conflicting information.

What does it feel like?

With attention we can get experiences of shifting awareness. We would be able to feel the difference between for example eyes open and closed. With memory we would be able to remember eyes open or closed, looking left or right. Previous levels of consciousness without attention would likely have continuous experience. The attentions means that the creature could ignore some stimulus in favour for others.

I suspect creatures with attention and memories would sleep, as the focus might be energy intensive and sleep allows energy use to be minimised and allows memories to be ingrained into reflexes from adaptive behaviours.

3.1.7 *Executive and Emotional*

Emotional agents are considered the first level of consciousness that actually exhibits conscious behaviours. From the "Theory of Mind" these agents exhibit complex emotions and complex behaviours to achieve goals.

Evolution of consciousness

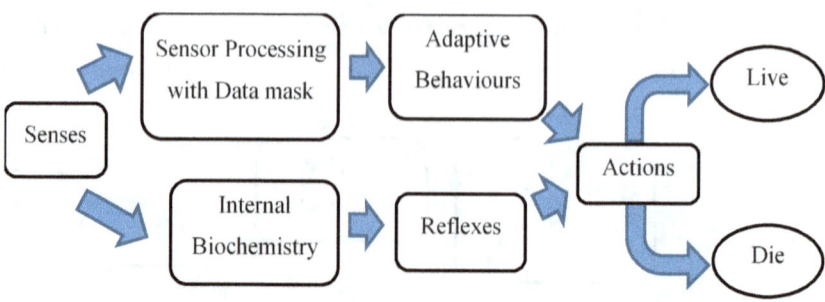

Figure 53 Executive

"Executive agents are defined by their ability to manage multiple tasks and "set shifting". The definition says "The emotional capabilities of living organisms at this level of consciousness seem to drive their actions." However, from the design, emotions would be a by-product of internal biochemistry and would be more or less integrated in Reflexes.

Executive control is more interesting in that it involves memory and focus. Once we have memory or at least "memory of senses" we can start considering a creature noetic. This means that they would be able to experience consciousness in the moment and will be able to recall past feelings, but likely not be able to consider themselves from an external perspective.

What does it feel like?

With Emotional responses we can start having associated feelings, feeling hot and a need to be active with anger, Cold with fright, warm and fuzzy when in love. The experience of living at this point becomes active and "multi flavoured". Meaning we can take actions that affect our emotions and that the emotions provide the variety in our experience.

3.1.8 Self-conscious

Self-conscious agents develop an understanding of self and the ability to recognise themselves in a mirror. These agents are often able to plan for the future.

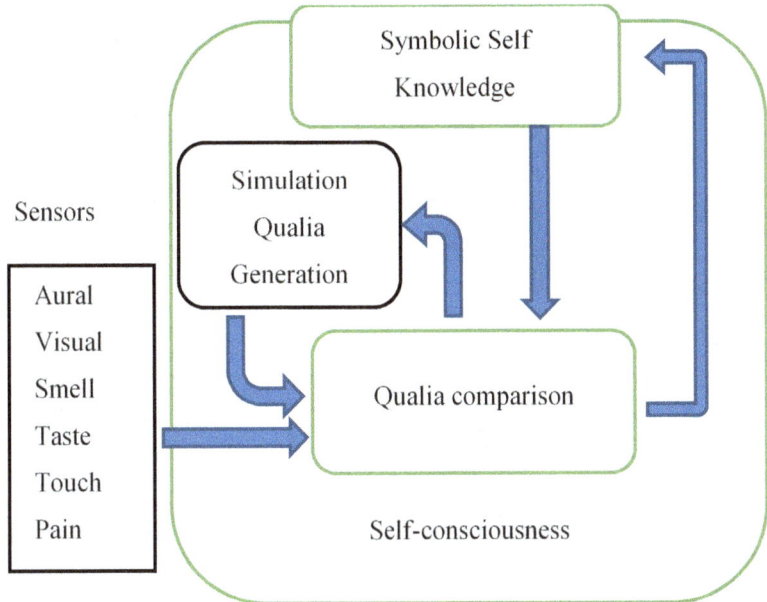

Figure 54 Detail of Self Simulation

For self-consciousness emerge the sensory data must be mapped to a model of the life form's body. This allows the life form to simulate potential actions and have the simulation mapped to incoming qualia from sensors.

At this point we can start considering a creature Autonoetic as it is able to consider the future and past. This is in contrast to previous levels where the model exists primarily for mapping sensors together. At this point the self-model can become active. This means that the creature can put itself in a simulation and consider future or past scenarios consciously.

What does it feel like?

Self-conscious a creature would have a full sense of self, being able to coordinate between sensors and have a good awareness between internal and external stimulus. Creatures would likely be able to recognise other creatures, but without empathy communication would likely be limited.

3.1.9 *Empathic*

Empathic agents are able to recognise emotions in other agents. At this level life forms are often capable of forming communities for self-preservation and create social structures.

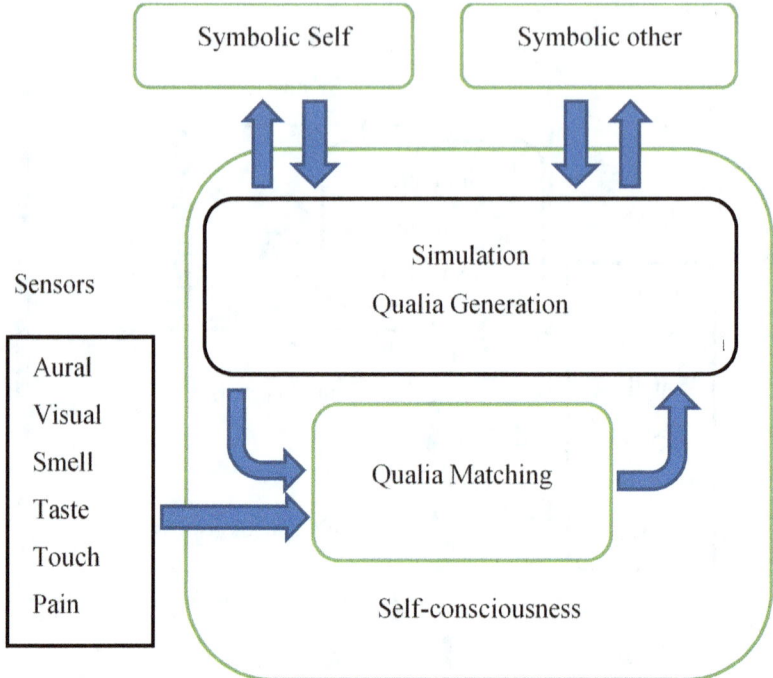

Figure 55 Detail of Self Simulation

Empathic consciousness develops from an awareness of other people. At least to be able to recognise the emotional expressions of other people. This introduces what I call a Symbolic other. Essentially a database of appropriate symbolic states relevant to other people.

What does it feel like?

Creatures with empathic capabilities would be able to recognise other individuals, so communication would likely be able to evolve here. Without complex conceptual understanding we probably would not see complex languages, but they may be able to be taught a limited vocabulary. To be able to use language they would likely have the capability to understand basic concepts.

3.1.10 Social

Social agents are able to recognise emotions in others and are able to develop strategies to take advantage of other agents and any social structure they are a part of. This level of intelligence is often called "Machiavellian" as agents are able to use cunning and deceit to complete their goals.

Evolution of consciousness

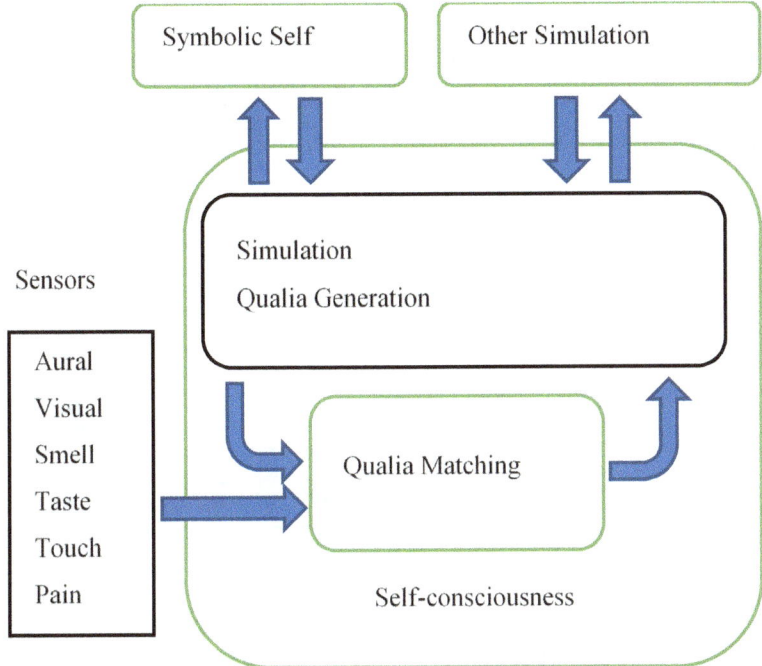

Figure 56 Detail of Self Simulation with social understanding

Social and Empathic consciousness develops from an awareness of other people. Recognising that others are also functional is part of the "Theory of Mind" described in Psychology. We can recognise that other people's knowledge and emotions change, and their actions may vary depending on their emotions and knowledge state.

I suspect that we use our own consciousness as a basis for estimating others actions. As such this results in our tendency to expect other people to make the same decision as ourselves in the same circumstance. This is called the "false consensus effect" and has been noted by psychology researchers [90]–[92].

What does it feel like?

Social creatures are likely to be able to communicate in many ways between individuals, not necessarily with complex language, but certainly vocally and with gestures. These creatures would likely be able to have a reasonable conceptual understanding.

3.1.11 *Human-like*

Human-like agents are capable of developing complex social systems, developing writing and using technology to modify their environment.

The human-like consciousness model is exactly what we have from the full design of consciousness. The model must allow all behaviours we recognise in each other. What you should notice from the previous sections in this chapter is that there is a development sequence. My guess is that this represents an evolution sequence from simple organisms to the behaviours we currently see in humans.

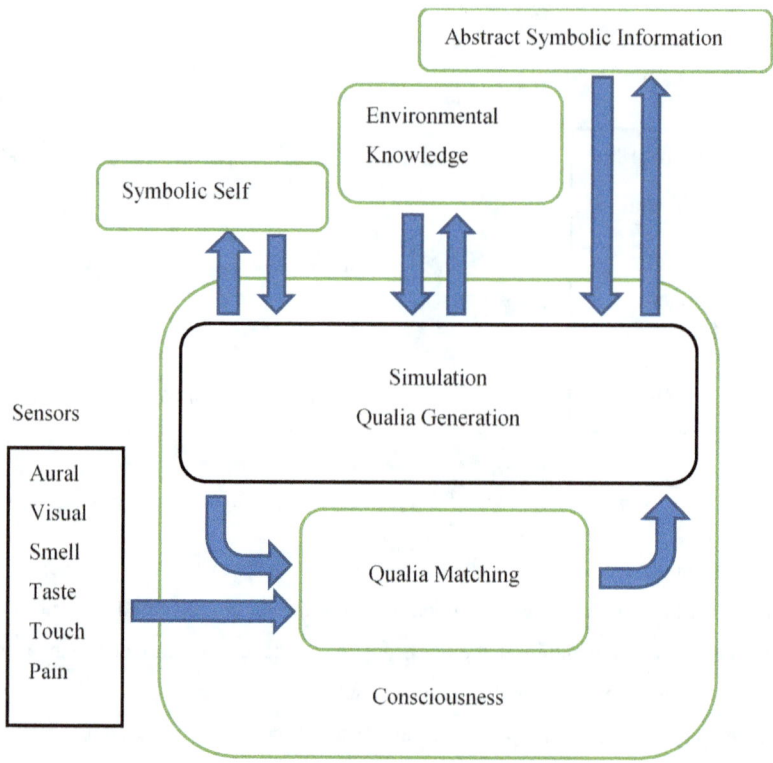

Figure 57 Detail of Self Simulation

The main features of human level consciousness is the self-model/awareness and the ability to model the world around us and make "conscious" decisions based on an internal simulation. Humans are considered Autonoetic, but I would argue here that we have at least had the capability for autonoeticism for several level on consciousness. At this point though we would be consider experts at autonoeticism.

What does it feel like?

If you are reading this you are *{Hopefully}* obviously aware of what human experience is like. We are capable of complex language, abstract thought combined with emotion.

{Unless you are an alien reading this from way in the future ...in which case: Hi from the past.}

3.1.12 Super-conscious

The Conscale *{and many science fiction stories}* consider the idea of "super human" consciousness. We can extrapolate from the basic human model and come up with at least one way to become "super" human. Super-conscious agents are defined as able to "synchronise and coordinate multiple streams of consciousness". This is the idea of a hive mind, where a single mind "has control" or at least coordinates information between individuals. We can model this idea as shown in Figure 58.

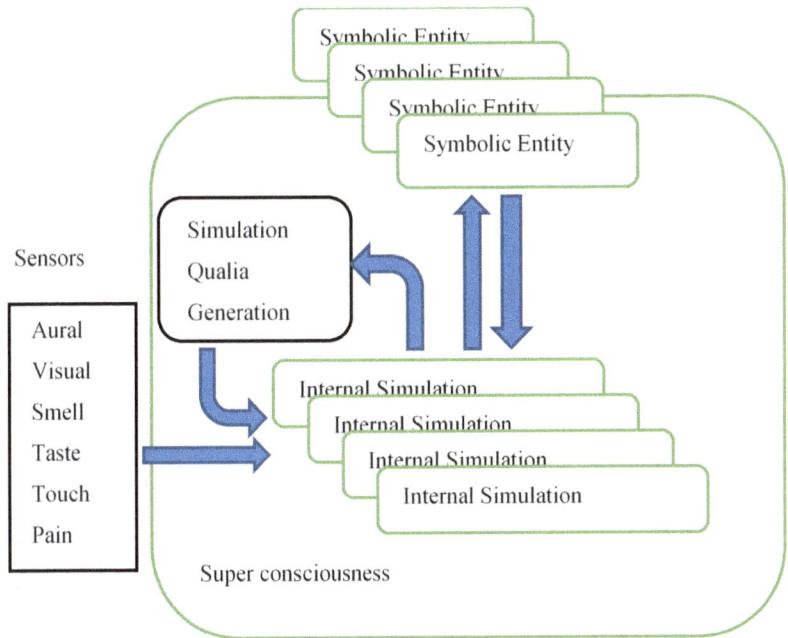

Figure 58 Hive Mind Model

This would require multiple individuals being simulated at the same time and potentially their interactions. This gets far more complex and requires linear complexity for each entity simulated and an n^2 (complexity or worse) for dealing with the interactions. As such, it is expected that a super consciousness would require significantly more processing power than the individuals themselves.

What does it feel like?

My best guess is that this would feel dissociated. Being aware of multiple points of view at the same time would potentially allow the entity to develop a "map view" to help relate the different points of view, or perhaps something like a multi-camera security feed. This would allow coordination between the individual entities. However, with such a dislocated intelligence we might find that the individual feelings could become less important. On the other hand if all of the individuals in the group consciousness share the same feelings it could cause an intensity of emotion throughout the group.

3.1.13   Conscale Summary

As with other aspects of this design other researcher may not completely agree with these descriptions. However, what we have shown here is that this design can be modified to address each level of consciousness with a coherent explanation. This again is something that gives me some confidence that this is a sensible design.

3.2   Important Features of development

At many points during our evolution "features" developed that enabled the expansion of diversity of species we see on earth. I suspect that the development of each of these features was followed by a large growth in the number of species present in the environment.

3.2.1   Instinctive

Connecting input sensors to outputs, gives creatures much more control over their actions and allows them to be much more efficient and effective especially at tasks such as hunting and avoiding being hunted.

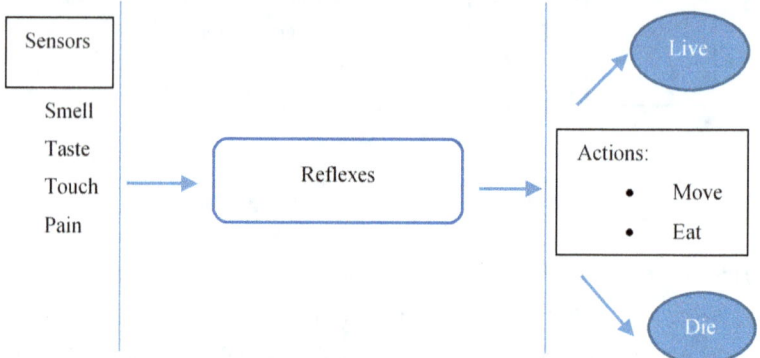

Figure 59 Process Diagram of a basic lifeform

Instinctive creatures, could be considered similar to Reactive creatures on the conscale.

3.2.2 *Predator – Prey arms race*

Predators and prey show up very early in the evolutionary tree. Down to jellyfish and even microbes we can find examples of life eating other life forms. Probably the first would be between animals and plants, but once this race is started it would likely cause creatures to evolve rapidly. My guess is that the predator prey arms race would start very early in the Conscale and would enable the development of noetic experiences quite quickly.

With multiple complex sensors a lifeform would probably need some way to connect the sensory information. As such, once lifeforms have multiple sensors and some memory I would expect concepts to begin to form to connect the senses. That said there are probably only two concepts: Food and Predator.

I suspect Prey is more likely to become conscious before predators. A Predator simply needs to get close and be faster/stronger/poisonous. As such, they can operate based on one concept: move towards food and eat it.

Prey has a second concept to look for: is this other thing attempting to eat me. Prey has to examine an environment and determine if something is attempting to kill it. This causes hesitation as the prey will evaluate its surroundings i.e. it will stop and think.

The more complex life forms become the more the features that prey can use to escape its predators. The more the prey tried to escape the more predators need to adapt to find and hunt their prey. Which give an arms race. There are many examples of how this leads to a lot of acceleration in evolution. In these situations we see complimentary features between predators and prey develop and eventually diversify into different strategies and producing many different species that use different strategies to hunt prey and avoid predators.

One of the big ones is memory and learning once we know we have a few things we need to run away from and a few things we can approach and eat, we need to remember them. Our memories give rise to many complex behaviours. The difficulty then becomes how to coordinate these actions, the more complex the actions the more complex the sensors we need to determine if they are necessary.

3.2.3 *Complex sensors*

In terms of the Conscale, creatures with complex sensors would exist in a wide range between Reactive and Adaptive levels. Simple sensors such as touch, temperature and pressure require the development of a nervous system. More complex sensors such as eyes, ears and smell require a brain to receive and process information. So complex sensors would likely give the capabilities of Adaptive and even Attentional or executive creatures.

The more complicated the situations creatures find themselves in the more complicated the sensors they use to examine their environment. We can imagine the how a predator prey arms race could cause sensors to become much more complex. The better a predator can detect its prey the more efficiently it can hunt down its prey. However, the better a prey creature can detect a predator and hide from it the more likely the prey creature will survive. So we end up with an arms race, where predator and prey motivate each other to develop new sensors, and techniques to hunt and avoid each other.

Complex sensory inputs include Eyes, Ears, and touch. These sensors provide a mapping of the world around us, the complexity of the sensor in general gives rise to more detailed information being collected from the world around us. The development of the sensors most likely also corresponds with a rise in the intellect to manage the incoming data producing intelligence the enables complex actions including Hunting, Hiding, and Ambushing. These actions require knowing the environment, how to move through it. The variation in sensors gives a rise to a variation in survival strategies which results in the huge diversity of species of creatures we see in the world.

The complex sensors once coordinated allow us to build a model of ourselves and the world around us. Once this model is effective we can make more and more complex actions.

3.2.4   Self-model

We know that people at least can be conscious of themselves. An associate behaviour is being able to recognise oneself in a mirror. This means we need to have an internal model of ourselves that is matched to incoming sensory day. We usually know where our limbs are so if we look in a mirror we can see that our mirror image performs the same movements that we do.

In biology one can image how this model is developed. When we are born we need to develop an understanding of our own body, realising that we have two arms and two legs. We also see this behaviour in other animals, a puppy chasing its own tail. We can think of the puppy's internal dialog:

"What is that thing moving behind me?"

"I should catch it"

"Got it I wonder what it tastes like"

"Bite it"

"Ouch"

"Oh that's MY tail"

A simple lesson, but this demonstrates a biological entity learning about themselves and adding to their self-model. The puppy initially did not include its tail in its self-model. Dogs have a chase reflex where they will run after any object moving away from them, so their tail moving away from their own field of vision attracts their attention, but if they do not realise the tail is part of themselves they still chase it. When they finally get hold of their tail (or if) they realise that they can feel themselves biting their own tail. The pain signals would give them the ability to connect their own self model with a tail to the visual and physical stimulus.

Most likely the first sensors developed would be nerves for touch or pain. Once you have them you can detect different aspects about the environment you are in and if any part of yourself was damaged. This allows creatures to avoid significant harm and learn of threats in their environment.

This probably developed as soon as we developed multiple sensors. We needed to figure out where the sensors were pointing, coordinate the inputs and build a self-model as well as a better world model. I suspect we can think of this as the starting point of consciousness. Essentially the self-model is a map of our sensory inputs that we can use to predict ourselves and the effect of actions on the external world.

In comparison to the Conscale, creatures with a self-model would start as soon as sensors need to be coordinated, so it would likely develop in parallel with the complexity of sensors. As the sensors become more complex and sensitive, creatures' sense of self would improve along with the ability to distinguish between internal and external stimulus. My feeling is that once we have a self-model essentially the lights go on and creatures become conscious at least at an anoetic level.

3.2.5 **Sexual Reproduction**

Sexual reproduction give another action to the list: Find a mate. This means that things become much more complicate as we now have 3 actions to choose between

Action list:

1. Eat
2. Run away
3. Find a mate

Sex results in the need for creatures to interact more. Which means they need to start developing social "skills". Due to the differences in the energy requirements by each sex different and often conflicting behaviours emerge. These are well covered by Dawkins and many other biologists. Suffice to say, even with males vs females we end up with various tactics developing to maximise the chances that offspring survive while minimising the energy used.

Sexual reproduction allowed changes to be made in species faster than from a single parent by exchanging genes between a pair of successful parents. This means that changes did not have to be random between each generation. The parents' genes could be mixed which allows creatures to explore the available "genetic space" without having to be waste effort on changing patterns required to survive, as occurs with purely random genetic drift.

3.2.6 *Social animals*

Social animals can rely on a group to help themselves survive. Non-social animals end up competing for the same resources, whereas social animals can cooperate to distribute resources more effectively.

Being "social" requires the development of an "Other Simulation". This allows us to predict what others might know, think and feel. It allowed us to transmit knowledge between individuals and gives the ability to cooperate.

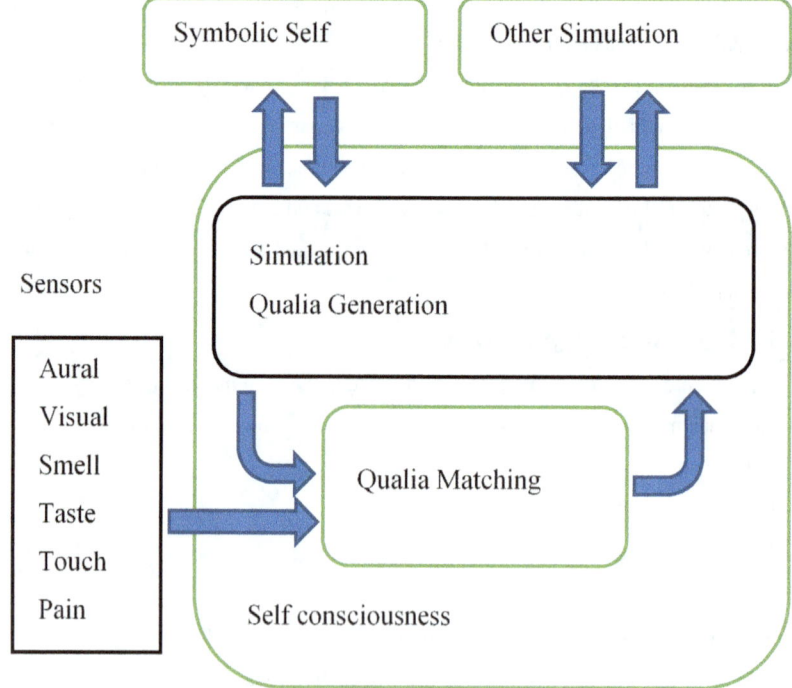

Figure 60 Detail of Self Simulation with social understanding

Ultimately this has led to the development of language, writing, and from that culture. This has allowed humans to preserve knowledge externally, meaning that individuals no longer need to re-learn everything the previous generation found out from scratch. We can simply read about their successes and failures, then build on them.

In itself being social I suspect requires individuals to be conscious beforehand at least at a noetic level and most likely enabling the development of autonoetic consciousness. They needed to be aware of themselves and having a self-model allows one to simulate another even imperfectly. As mentioned previously using your own self simulation model produces the "false consensus effect", which is problematic in itself, but is also a kind of confirmation that this is a sensible idea for how our own consciousness operates.

3.2.7   *Empathy*

Proverb "Before you judge a man, walk a mile in his shoes."
Steve Martin "That way you are a mile away and have his shoes"

We define empathy as the ability to understand and share the feelings of another. In practice with the design we use a "self-simulation" to determine how the world interacts with the "self". From this a consciousness can determine how it "feels". With some amount of lateral thinking it is possible to apply a self-simulation to another "person" (or entity). This essentially give the consciousness a capability to determine what it would do in someone else's place.in a sense empathy.

The other factor of empath is the feature of "mirror" neurons in people and some animals that are triggered in response to someone else's facial expressions among other stimulus. This is noted in babies especially when forming bonds with parents. It can be reasoned that this has developed in an evolutionary sense to assist the child surviving to adulthood by ensuring that the parent cares for the child.

3.2.8   *Intelligence*

Humans ability to handle abstract concepts such as Mathematics has led to our ability to modify our own environment and ultimately to survive and expand our population across the entire planet. Our ability to generate social constructs such as religion and culture have allowed us to form cohesive *{well, comparatively cohesive}* societies without having individuals continually competing for resources.

Arguably the first stages of intelligence come from the predator prey arms race and complex sensors. The more complexity our intellects can handle the better we can survive in a hostile environment. From this we can argue that intelligence is another major factor that allowed species to develop. The different survival strategies allow different species to specialise in different environments, which allows some species to find and populate areas of the world that are less hostile. With the predator prey race this will eventually bring predators and the cycle begins again.

3.3   When did consciousness evolve?

From reviewing some animals in the context of evolution I would guess that consciousness evolved in the predator- prey arms race. This arms race is often less about physical characteristics such as strength and speed and more about strategy and tactics.

Many creatures may naturally evolve to seek particular environments, insects often seek dark moist places to live in. These behaviours can be simulated in robots with simple light and moisture detecting sensors. These robots are just reflex agents simply responding directly to sensors with inbuilt reflex actions: move to where it is darkest, move to where it is warm and wet.

To get from reflex agents to conscious entities requires at least one more component. I suspect we can think of the self-model as the starting point of consciousness. The predator prey arms race is where or why the self-model evolved. The self-model is a map of our sensory inputs that we can use to predict ourselves and the effect of actions on the external world. I expect that one way we could consider the complexity of consciousness of a creature by its ability to map itself and its external world. The more features, both internal features (eyes, legs etc.) and external features of the world, a creature can map and be aware of the higher level of conscious a creature could be considered.

Consciousness as shown in the design requires sensory input, memory and the ability to simulate probably both itself and many things to do with a life forms surroundings. A conscious entity needs to be aware of itself and probably its surroundings.

When we consider why we can look at this problem from an evolutionary perspective. How does consciousness allow a lifeform to survive? My guess is tactics, the basis of our intelligence. We have two ingrained survival instincts: Fight or Flight. These tactics themselves are not enough to determine if something is conscious. However, an understanding of when to use which strategy, I suspect, may lead to consciousness. Consider a deer in a forest being hunted by a large cat. When the deer initially spots the cat it has to think: Am I strong enough to fight this thing or am I fast enough to get away. If I fight how do I attack? If I run, which direction? These questions may not be expressed in language, but if you watch this situation carefully you can see the deer considering its situation, looking at the cat to see if it is just a cub, feinting or sidestepping to get in a better position. These decisions require an active thought process about the situation. An active thought process requires awareness of a situation and self. Together these suggest consciousness.

From this we can examine consciousness to see if we can find a way to determine if something is conscious. These may not be a complete list or a completely accurate way of determining consciousness, but given the design it is a starting point of where we can say something is or is not conscious. We have already established that a lifeform that uses or creates structured language is probably conscious. One testable idea for consciousness is to see if the lifeform in question can change its mind.

However, this may not be true for a computer system or a robot, but we will get to that in Chapter 6.

4 Examining Biological Consciousness

OK we have described the "design" and shown how it could have been generated through an evolutionary process.
In what ways can it be tested to explain what consciousness is and how it makes "us".

To test this idea part of what we can do is examine the similarities and differences between what we do know is conscious in biology i.e. us, and this design. While it probably isn't going to be conclusive, hopefully it will give future researchers a few ideas about how consciousness operates and how to test the design of consciousness and other ideas further. The goal is to examine the structure and operation of the brain and to see if and where particular processes might occur.

If you are interested in learning more about the brain you can find far more accurate descriptions in these books:

- R. Carter, The human brain book: An illustrated guide to its structure, function, and disorders. Penguin, 2019.
- T. Vanderah, The human brain: an introduction to its functional anatomy. Philadelphia: Elsevier, 2016.
- M. S. Greenberg, *Handbook of Neurosurgery. 7 th.* Thieme New York:, 2006.
- D. Wolpert, K. Pearson, C. Ghez, and E. Kandel, "Principles of Neural Science," *The Organization and Planning of Movement. 5th ed.* New York: McGraw-Hill, pp. 475–97, 2013. [93]

There are also many online descriptions including:

- https://www.healthline.com/
- https://brainmadesimple.com/
- https://www.brainfacts.org/
- https://www.britannica.com/browse/Anatomy-Physiology
- https://www.neuroscientificallychallenged.com

… and many others.

Most of the descriptions of the brain here are restated from these and other online resources. I am not an expert in the structure of the brain, hence the descriptions may not be clinically accurate, but hopefully it will be accurate enough to get some idea of how the design relates to what we find in the structure of the brain. So any hypothesis presented in this section I am perfectly happy to admit is wrong given someone else's expertise.

4.1 The Brain

We know consciousness exists and most likely is a phenomena inside the brain. So are there many comparisons we can make between the design and what we know about how the brain produces consciousness.
{Brain Surgeon: Braaaains! What?!?!?! NO I AM NOT A ZOMBIE}
{The Brain: Please don't eat me!}

In the following section I have summarised the structure and associated function of the human brain. I ask some forgiveness if I get some of this a little off as I am not a neurobiologist/surgeon/brain-specialist. The idea is to show that I have done some reading and have at least a basic understanding of the human brain in the context of human consciousness. The up side is that I am a human (surprising to some) and we all have some experience with consciousness.

The scientific community has not yet established the mechanisms in the brain that allow consciousness to exist. However, we have made great strides recently in understanding what we can see through the use of Functional Magnetic Resonance Imaging (FMRI), Positron Emission Tomography (PET) and other technologies.

The outer layer of the brain is referred to as the Cerebral cortex. This includes four lobes: frontal, parietal, temporal and occipital. The internal structures of the brain include the Corpus Callosum and the Cerebellum. Leading up from the brain stem are the medulla and pons which connect to the Corpus callosum.

Hypothesis:

My suspicion is that the Cerebral cortex is actually our long term memory most likely along with architecture that allows it to analyse and reproduce such signals. This analysis and reproduction allows concepts to be generated in working memory and used to simulate anything humans think of to perform planning tasks and predict outcomes.

We have also noted through a large body of testing that different areas/lobes of the cerebellum relate to different types of information. As shown by other researchers Wernicke and Broca's areas are associated with language. Wilder Penfield is known for mapping the human brain in the initial days of brain surgery. He is often quoted (from "In conversation with Denis Brian", reported in "Genius Talk", 1995) about probing a patient's brain with an electrode and managing to activate an area in the patient's brain so that she heard a melody. "I was so astonished I re-stimulated the same spot 30 times. Each time she heard the same melody". Other than being interesting in itself, this shows that memories can be accessed through stimulating locations in the brain.

If you examine a CPU as shown in section 4.3 we find that the largest part of any CPU is the memory system. It would not surprise me if a similar relationship exists with biological brains. The design uses multiple "Databases" for different purposes. This suggests that it is a sensible strategy to have a large amount of storage accessible with specific areas being separated for specific purposes. Our memory is also somewhat relational, where for example colours or smells can allow us to access memories. This suggests we have system where a signal in one area of the brain could trigger a response in another area. The cerebral cortex is a continuous surface which may allow connections between different areas of the brain, which would give us the best of both worlds where areas storing similar information could be grouped together. Finally the data stored in our memory could be either information or in effect "programs" for specific purposes. As with computers programs can be stored on hard drives or in memory as data. So in general terms this means I am using the term "memory" somewhat loosely.

The Hippocampus is related to managing memory however memories are not stored in the hippocampus. The structure and operation of the Cerebral cortex may also explain why we have never found a specific area of the brain that stores memories. The memories are distributed widely across the brain and the purpose is as much an analytical tool to examine sensory information and potentially reconstruct imagery from memories used when attempting to simulate the world around us. Secondly these memories may be "active", equivalent to storing a program as opposed to say an image on a hard-drive. The hippocampus may also be storing instructions of how to perform actions or perform analysis as well as static data, names, places, faces all the way up to abstract concepts.

This hypothesis is supported by the observation from Dehaene[94] of specific neuron groups for recognising specific people i.e. we have a Bill Clinton detector.

4.1.1 Overview of Structures

The brain can be split into many regions by structure and function. Figure 61 show an overview of these structures.

Examining Biological Consciousness

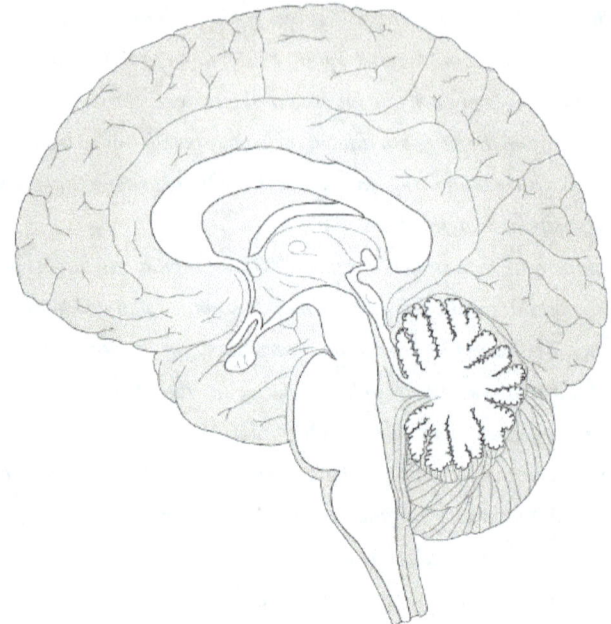

Figure 61 Structures of the Brain

4.1.2   Neurons

One of the features recognised by Edelman, Hierarchical Predictive coding and Baars research is that neurons can use the same links to transmit and receive signals. This is referred to as re-entrance or re-entrant signalling. The bi-directional nature of neurons would simplify many functions in processing as signals can be retransmitted back along the same neural pathways.

Hypothesis:

There may be up to 3 purposes for the re-entrant nature of neurons:

1. The re-entrance property might be useful in reprograming connection values between neurons in a backpropagation system. In other words, they might be part of the system that helps us reprogram our neural networks while we sleep.

2. It may be part of an error correction system that allows neurons to be replaced if there are not functioning correctly. The re-entrant property could be part of a self-testing system for maintaining the connections.

3. The bi-directional nature of the neuron clusters may simplify the communication structure. If we consider how neural networks operate they send a set of signals through the network, it could be a diverging network that for examples takes an image and matches it to a template for object recognition. So in this case we might see the image being "uploaded" to the neural net and a return signal returned showing which template was identified through a single re-entrant neural connection.

4.1.3   Brain Stem and Autonomic Centres

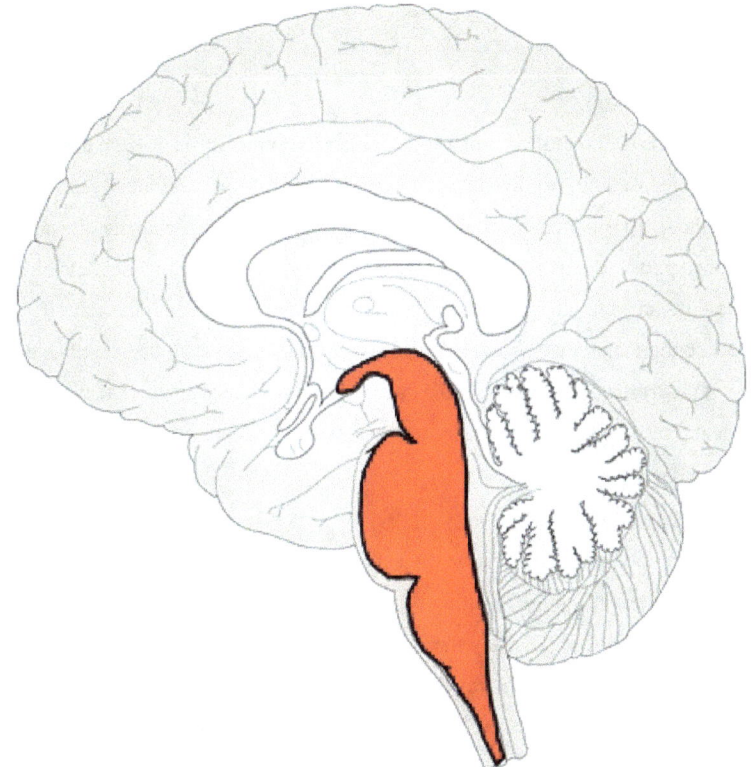

Figure 62 The Brainstem

The brainstem is the lowest part of the brain. It connects the spinal cord to the interior sections of the brain. In the human brain the brainstem includes the midbrain, the pons, and the medulla oblongata.

The brainstem is a very small component of the brain, making up only around 2.6 percent of its total weight. The autonomic nervous system controls internal body processes such as the following:

- Blood pressure
- Heart and breathing rates
- Body temperature

- Digestion
- Metabolism (thus affecting body weight)
- The balance of water and electrolytes (such as sodium and calcium)
- The production of body fluids (saliva, sweat, and tears)
- Urination
- Defecation
- Sexual response

Many organs are controlled primarily by either the sympathetic or the parasympathetic division. Sometimes the two divisions have opposite effects on the same organ. For example, the sympathetic division increases blood pressure, and the parasympathetic division decreases it. These two parts systems act in concert as a control system detecting any deviation from optimal values and applying changes in the system to stabilise the internal states.

The autonomic nervous system is the part of the nervous system that supplies the internal organs with nutrients and control signals either via nerve signals or internal chemistry (hormones). The autonomic nervous system has two main divisions:

- Sympathetic
- Parasympathetic

After the autonomic nervous system receives information about the body and external environment, it responds by stimulating body processes, usually through the sympathetic division, or inhibiting them, usually through the parasympathetic division.

An autonomic nerve pathway involves two nerve cells. One cell is located in the brain stem or spinal cord. It is connected by nerve fibres to the other cell, which is located in a cluster of nerve cells (called an autonomic ganglion). Nerve fibres from these ganglia connect with internal organs. Most of the ganglia for the sympathetic division are located just outside the spinal cord on both sides of it. The ganglia for the parasympathetic division are located near or in the organs they connect with.

The brain stem also provides the main motor and sensory nerve supply to the face and neck via the cranial nerves. Other roles include the regulation of the central nervous system and the body's sleep cycle. In reference to the design the brain stem are a part of internal biochemistry and reflex actions.

It is also of prime importance in the conveyance of motor and sensory pathways from the rest of the brain to the body, and from the body back to the brain. These pathways include the corticospinal tract (motor function), the dorsal column-medial lemniscus pathway (fine touch, vibration sensation, and proprioception), and the spinothalamic tract (pain, temperature, itch, and crude touch).

Medical conditions involving the brains stem demonstrate the function of these components and show how essential they are for life and consciousness to occur. Diseases of the brainstem can result in abnormalities in the function of cranial nerves that may lead to visual disturbances, pupil abnormalities, changes in sensation, muscle weakness, hearing problems, vertigo, swallowing and speech difficulty, voice change, and co-ordination problems.

Brainstem death is one condition used to make the decision of when death can be declared, to stop ventilation of somebody who could not otherwise sustain life. Other determining factors are that the patient is irreversibly unconscious and incapable of breathing unaided.

Hypothesis:

In the design these functions are defined in Sensors and Sensory Processing to the brain and Reflexes or motor control to the rest of the body. In Computer terms this is that main I/O bus.

4.1.3.1 Midbrain

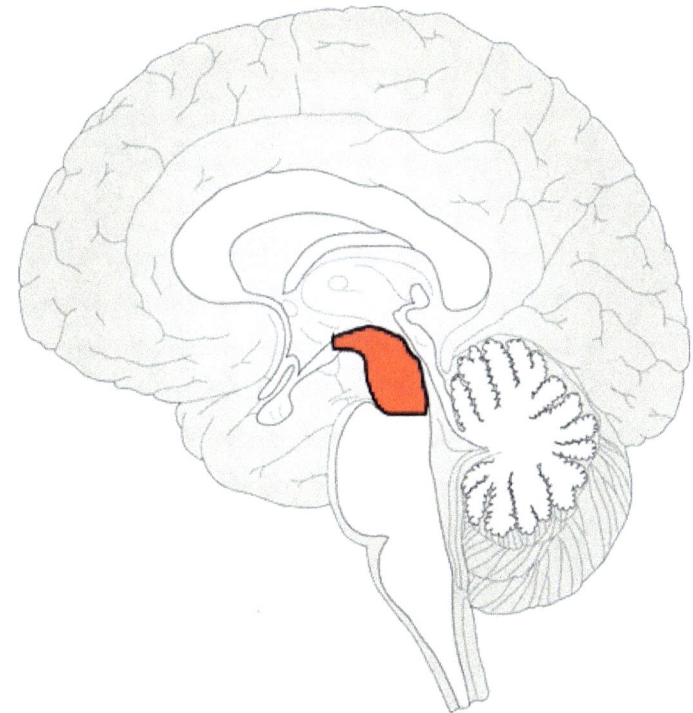

Figure 63 Midbrain

The midbrain is positioned above the pons, at the very top of the brainstem, directly underneath the cerebellum. All neuronal transmissions that pass through the body to and/or from the brain pass through the midbrain. Four bumps on its posterior surface (the side that faces the back of the brain). Those bumps are indicative of the presence of four large underlying clusters of neurons; the upper pair are the superior colliculi and the lower pair are the inferior colliculi. The superior colliculi are thought to be involved in directing behavioural responses toward stimuli in the environment, and the inferior colliculi are best known for their role in auditory processing.

The midbrain is associated with the management of vision, hearing, motor control, sleep and wake cycles, alertness, temperature regulation and even pain suppression. The midbrain is closely related to the movement of the eye and pupil construction.

Hypothesis:

Given the complexity the midbrain may be closest related to Sensory Processing in the design, specifically focus or data masking as eye motor control and focus related to concentrating on a particular distances.

4.1.3.2   Pons

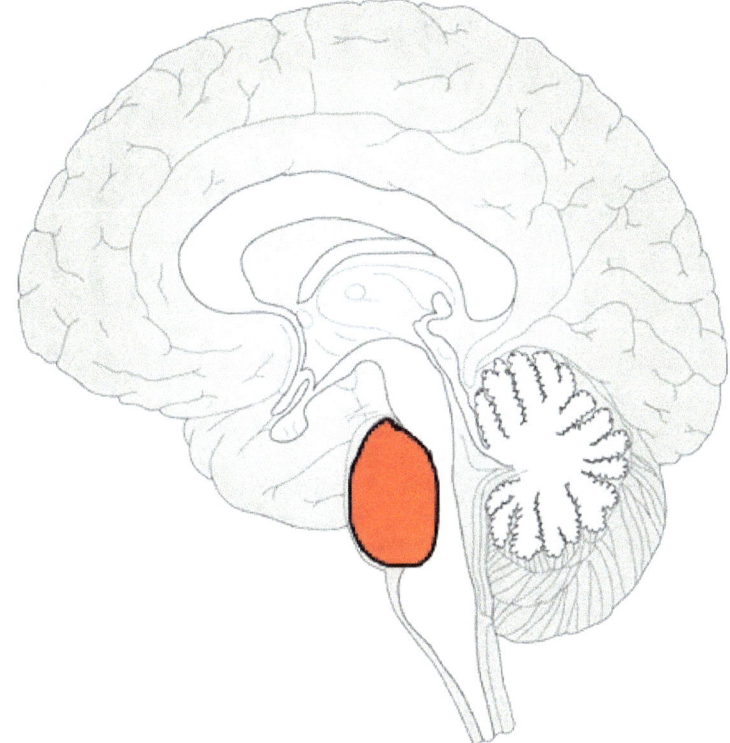

Figure 64 Pons

The pons is a 2.5 cm long portion of the brain stem, located above the medulla oblongata and below the midbrain. It is a bridge between various parts of the nervous system, including the cerebellum and cerebrum.

As part of the brain stem, the pons impacts several automatic functions necessary for life including the intensity of breathing, and a section of the upper pons decreases the depth and frequency of breaths. The pons contains cholinergic, monoaminergic, and serotoninergic neurons which are involved in control of attention and sleep cycles.

4.1.3.3   Medulla oblongata

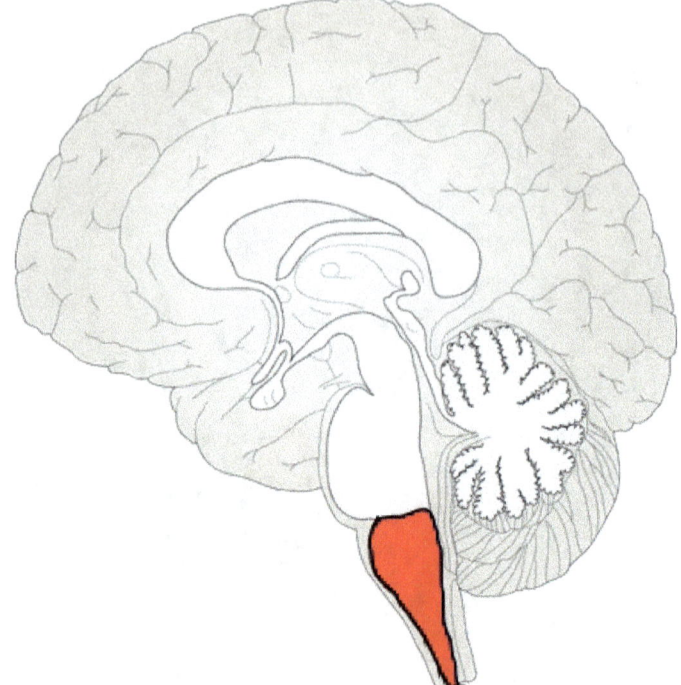

Figure 65 Medulla Oblongata

The medulla oblongata is located in the brain stem, in front of the cerebellum. This is a cone-shaped, neuronal mass in the hindbrain. It helps regulate breathing, heart and blood vessel function, digestion, sneezing, swallowing, control of movement, control of alertness and sleep and helps transfer messages between the spinal cord and the thalamus.

Injuries or diseases affecting the middle portion of the medulla may result in medial medullary syndrome, characterized by partial paralysis of the opposite side of the body, loss of the senses of touch and position, or partial paralysis of the tongue. Injuries or disease of the lateral medulla may cause lateral medullary syndrome, which is associated with a loss of pain and temperature sensations, loss of the gag reflex, difficulty in swallowing, vertigo, vomiting, or loss of coordination.

Hypothesis:

Given its proximity to the spinal cord and associated sensory information the Medulla oblongata could be associated with transferring sensory information and reflexes in the design.

4.1.3.4   Cerebral Ventricles

The brain ventricles are four cavities located within the brain that contain cerebral spinal fluid (CSF). When referred to altogether, the brain's ventricles are called the ventricular system. There are four distinct ventricular spaces: the lateral ventricles, the third ventricle, the cerebral aqueduct, and the fourth ventricle.

The first two lateral ventricles are C-shaped chambers found in the cerebral hemispheres (one in each hemisphere). They are connected to the third ventricle by an opening called the interventricular foramen. The third ventricle is a very narrow cavity that runs along the midline of the diencephalon; it communicates with the fourth ventricle via the cerebral aqueduct. The fourth ventricle is wedged between the cerebellum on one side and the brainstem on the other; it extends to, and is continuous with, the central canal of the spinal cord. The cerebral aqueduct is a narrow 15 mm conduit that allows for cerebrospinal fluid CSF to flow between the third ventricle and the fourth ventricle. Within each of them is a small structure called the choroid plexus. This is responsible for the production of cerebrospinal fluid, or CSF. CSF is a colourless fluid that performs many different functions, the most notable of which are the suspension of the brain within the skull and regulation of the neuronal environment surrounding the brain.

4.1.3.5   Reticular Formation

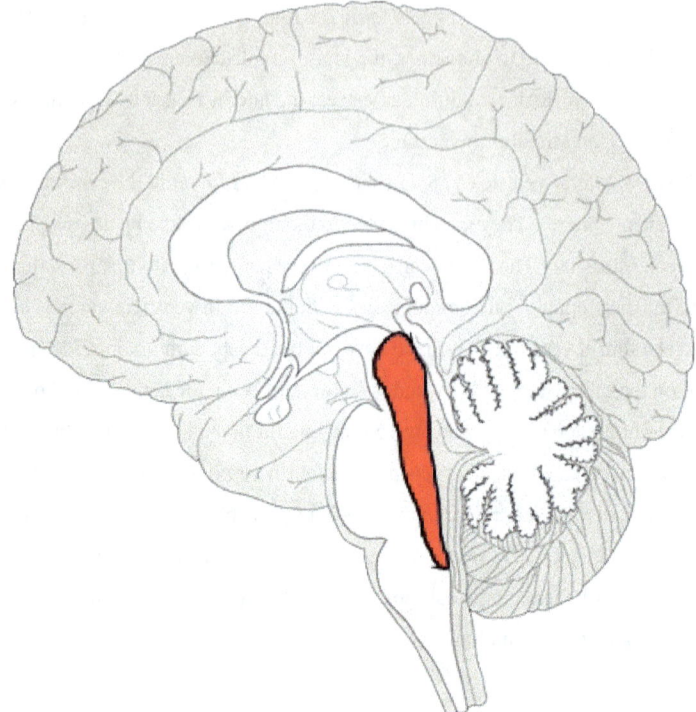

Figure 66 Reticular formation

The reticular formation is essentially the core of the brainstem, at the core of an area of the tegmentum. The tegmentum is a heterogeneous section of neural tissue that extends vertically through the brainstem, making up the portion of the brainstem that sits between the ventricles and surface structures like the basal pons and the pyramids of the medulla. The fibres that traverse the reticular formation give the region a net-like appearance, reticular means "net-like". The reticular formation contains several groups of cells that produce neurotransmitters including dopamine (ventral tegmental area), noradrenalin (locus ceruleus), serotonin (raphe nuclei), and acetylcholine (the pedunculopontine nucleus and laterodorsal tegmental nucleus). Neurotransmitters are produced in all of these areas and sent throughout the central nervous system to modulate sensory perception, motor activity, and behavioural responses.

As the reticular formation passes through the Midbrain, Pons and Medulla Oblongata it shares much of their functions. These functions include vision management, hearing, motor control, sleep and wake cycles, alertness, temperature regulation and pain suppression from the midbrain, autonomic breathing from the Pons, breathing, heart and blood vessel function, digestion, sneezing, swallowing, control of movement, control of alertness and sleep from the Medulla Oblongata.

The fibres that arise from these locations combine with other pathways that ascend to the cerebral cortex and thalamus to promote awareness. These pathways from the reticular formation must be functional for normal attentional abilities and sleep-wake cycles to be preserved. Lesions to major pathways of the reticular activating system can thus impair consciousness, and severe damage can cause coma or a persistent vegetative state.

The reticular formation has been identified by Baars[19] as central to the role of consciousness. Considering it's central role in awareness and transmitting sensory data, this is a reasonable conclusion.

Hypothesis:

The reticular formation and the Pons are noted to be able to suppress pain signals. One of the features Baars[19] specifically mentions as well is the ability to "spotlight" specific stimulus and allows others to be ignored or suppressed. This is also A part of the Sensory processing in the design. This would simply require the same process for suppressing pain signals to be applied to other sensory stimulus.

4.1.4   The Cerebellum

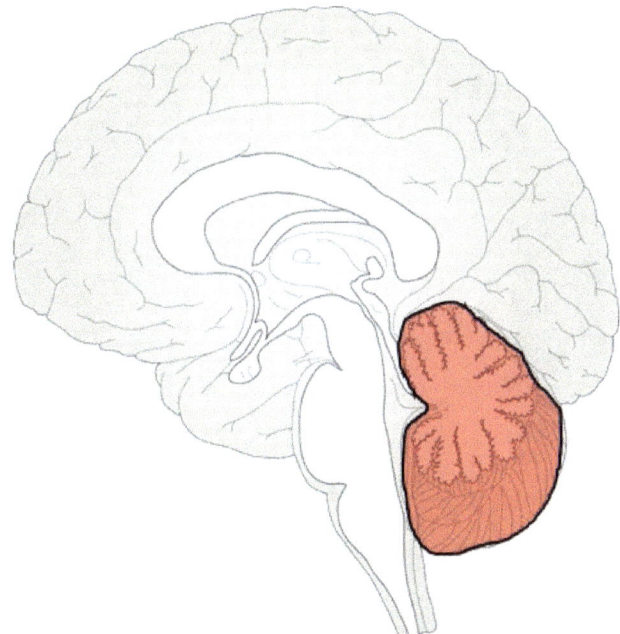

Figure 67 Cerebellum

Sometimes referred to as the "Little Brain," the cerebellum lies on top of the pons behind the brain stem. The cerebellum, including the vermis, is comprised of small lobes and receives information from the balance system of the inner ear, sensory nerves, and the auditory and visual systems. It is involved in the coordination of movements as well as motor learning.

The cerebellum makes up approximately 10 percent of the brain's total size, but it accounts for more than 50 percent of the total number of neurons located in the entire brain. This structure is associated with motor movement and control, it serves to modify these signals and make motor movements accurate and useful. For example, the cerebellum helps control posture, balance, and the coordination of voluntary movements. This allows different muscle groups in the body to act together and produce coordinated fluid movement and is even involved in speech.

Hypothesis:

This sounds to me like what we call in engineering a control system. Literally something that coordinates between sensors, motor control output and the desired final location. A control system aims to minimize the difference between desired output and measured/sensed output.

4.1.5   The Thalamus

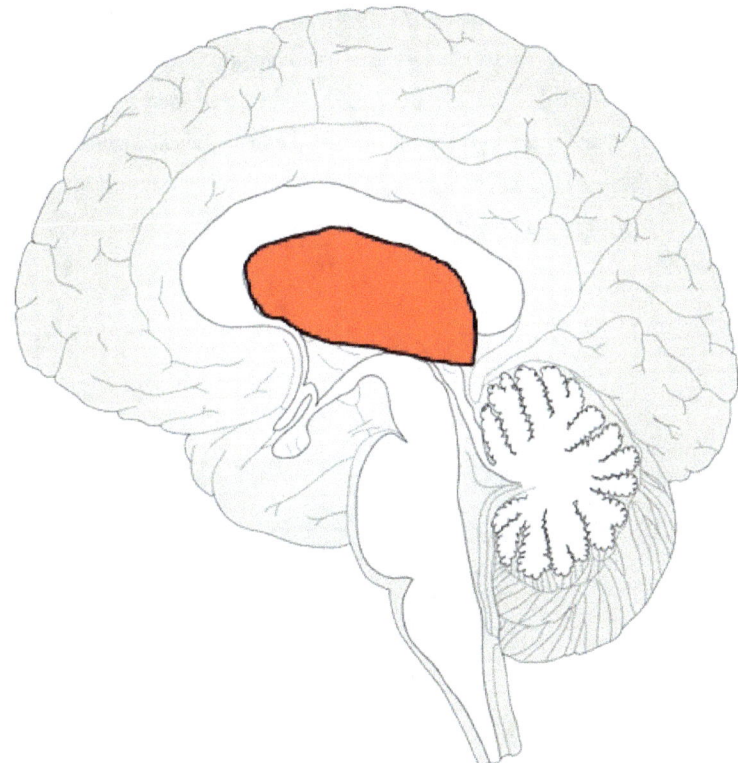

Figure 68 Thalamus

The Thalamus is located above the brainstem. It is close to the Reticular Formation at the top of the brain stem. It processes and transmits movement and sensory information to and from the cerebral cortex and Brainstem. The cerebral cortex also sends information to the thalamus, which then sends this information to other systems. The main function of the thalamus is to relay motor and sensory signals to the cerebral cortex. It also regulates sleep, alertness and wakefulness.

Hypothesis:

The Thalamus is also specifically mentioned by Baars[19] *As a structure that is involved in consciousness as it is also able to filter information flowing to and from the cerebral cortex.*

My hypothesis is that there is some structure in the brain that is used to simulate the external environment. It is possible that this is the thalamus, but it is difficult to determine if and where this process occurs. The thalamus is a possibility as it has a central role in transmitting data to and from the Cerebral cortex where our memories could be and sensory information is received from the body. The information received from the Cerebral cortex may be used to both update memories and used as a recognition system to build the simulation. If the Cerebral cortex also supplies the rules for predicting outcomes this could form a coherent picture of the world in the thalamus and be matched with simulation data again in the thalamus. The matching process may also occur in areas of the cerebral cortex.

4.1.6 The Basal Ganglia

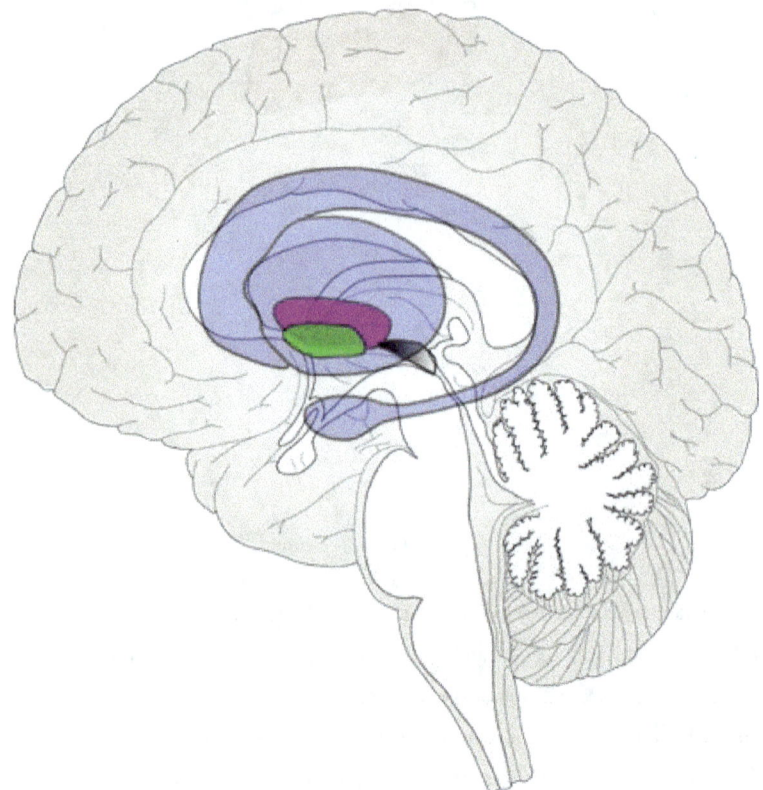

Figure 69 Basal Ganglia

The basal ganglia are a group of large nuclei that partially surround the thalamus. These nuclei are important in the control of movement. The following are functions have been established by many researchers [95]–[98]:

- Planning and modulation of movement pathways
- Reward processing and motivation
- Decision making

- Working memory
- Eye movements

Moreover, the basal nuclei use proprioceptive feedback from the periphery to compare the movement patterns generated by the cerebral cortex with the actual movement, so that the movement is subject to ongoing refinement by a continuous control mechanism.

The basal ganglia have been shown to play an important role in motivation. Considering that the basal ganglia circuits are influenced heavily by extracellular dopamine, high levels of it have been linked to satiated "euphoria", medium levels with seeking and low with aversion. The activation of the basal nuclei pathway that causes the disinhibition of the thalamus, leads to activation of the prefrontal cortex and ventral striatum. There is also evidence that other basal ganglia structures including the globus pallidus, pars medialis and subthalamic nucleus are involved in reward processing.

Regarding memory, the same structures in the prefrontal cortex are shown to be involved in the memory gates and focus. By using the basal ganglia's direct and indirect pathways as a relay between the input information from surroundings to the cerebral structures involved in memory storage.

The caudate nucleus is part of the basal ganglia, an elongated C-shaped nucleus that lies anterior to the thalamus, just lateral to the lateral ventricles and medial to the internal capsule. It is involved in many tasks, such as memory, goal-pursuit, learning, language processing, and emotions.

Hypothesis:

The Basal Ganglia sound like the required feed forward estimators required for estimating the next state in a simulation. They are shown to be related to planning decision, making and working memory, which are the central components of the simulation concept. As such, this may be a part of the simulation process and hence part of the consciousness loop.

4.1.7 **The Hypothalamus**

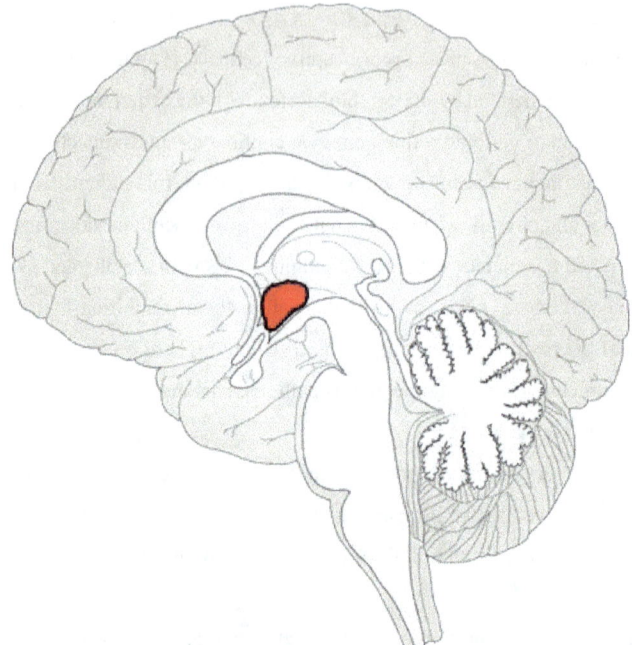

Figure 70 The Hypothalamus

The hypothalamus is a structure that lies along the base of the brain near the pituitary gland, between the Thalamus and midbrain. The hypothalamus is responsible for controlling hunger, thirst, emotions, body temperature regulation, and circadian rhythms. The hypothalamus also controls the pituitary gland by secreting hormones, which allows the body to achieve homeostasis and generally control the development of the body.

Primary hormones secreted by the hypothalamus include:

• Anti-diuretic hormone (ADH): This hormone increases water absorption into the blood by the kidneys.

• Corticotropin-releasing hormone (CRH): CRH sends a message to the anterior pituitary gland to stimulate the adrenal glands to release corticosteroids, which help regulate metabolism and immune response.

• Gonadotropin-releasing hormone (GnRH): GnRH stimulates the anterior pituitary to release follicle stimulating hormone (FSH) and luteinizing hormone (LH), which work together to ensure normal functioning of the ovaries and testes.

Examining Biological Consciousness

- Growth hormone-releasing hormone (GHRH) or growth hormone-inhibiting hormone (GHIH) (also known as somatostain): GHRH prompts the anterior pituitary to release growth hormone (GH); GHIH has the opposite effect. In children, GH is essential to maintaining a healthy body composition. In adults, it aids healthy bone and muscle mass and affects fat distribution.
- Oxytocin: Oxytocin is involved in a variety of processes, such as orgasm, the ability to trust, body temperature, sleep cycles, and the release of breast milk.
- Prolactin-releasing hormone (PRH) or prolactin-inhibiting hormone (PIH) (also known as dopamine): PRH prompts the anterior pituitary to stimulate breast milk production through the production of prolactin. Conversely, PIH inhibits prolactin, and thereby, milk production. Thyrotropin releasing hormone (TRH): TRH triggers the release of thyroid stimulating hormone (TSH), which stimulates release of thyroid hormones, which regulate metabolism, energy, and growth and development.

4.1.8 Cerebral cortex

The cerebral cortex is the outer covering of grey matter over the hemispheres. This is typically 2- 3 mm thick, covering the gyri and sulci. Certain cortical regions have somewhat simpler functions, termed the primary cortices. The cerebral cortex is usually classified into 4 main areas or lobes: Temporal, Frontal, Parietal and Occipital.

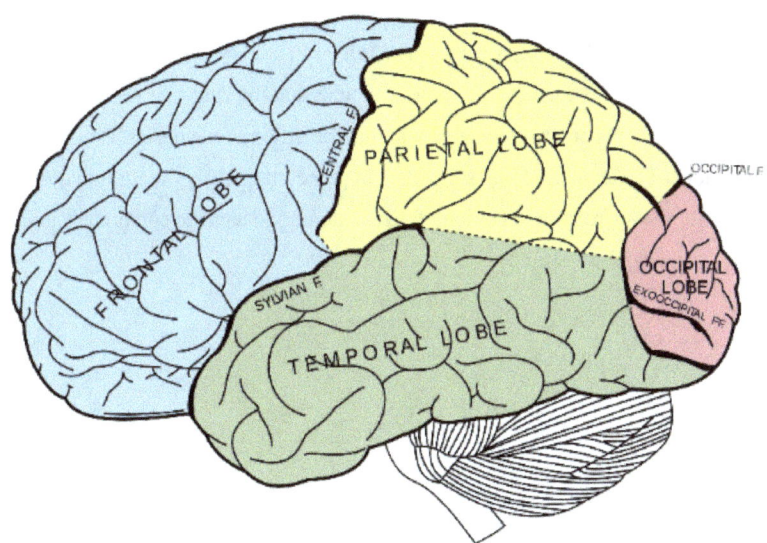

Figure 71 Major Lobes of the Cerebral cortex

4.1.8.1   Temporal Lobe

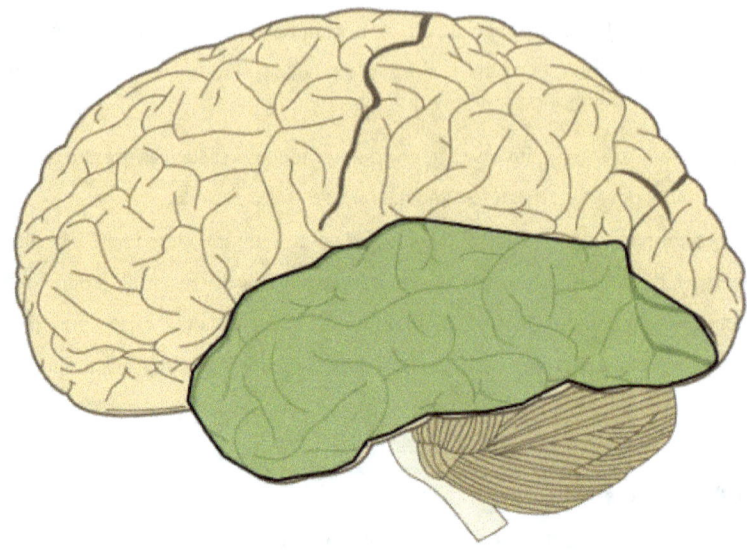

Figure 72 Temporal Lobe

The bottom section of the cerebral cortex is the temporal lobe. The temporal lobe is located beneath the lateral fissure on both cerebral hemispheres of the mammalian brain. "Temporal" refers to the head's temples. The primary auditory cortex is also located in the temporal lobe, which is used for interpreting sounds and the language we hear. The hippocampus, which heavily associated with the formation of memories, is also located in the temporal lobe. The temporal lobe is involved in processing sensory input into derived meanings for the appropriate retention of visual memory, language comprehension, and emotion association.

4.1.8.1.1 Amygdala

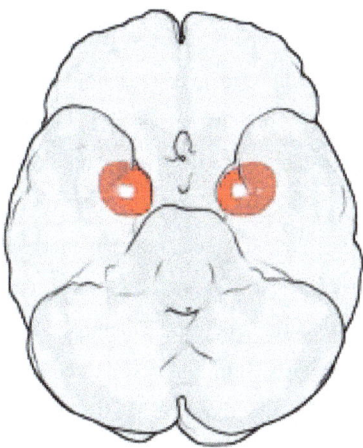

Figure 73 Amygdala

The amygdala is one of two almond-shaped clusters of nuclei located deep and medially within the temporal lobes of the brain in complex vertebrates, including humans. Shown to perform a primary role in the processing of memory, decision-making and emotional responses (including fear, anxiety, and aggression), the amygdalae are considered part of the limbic system. The term amygdala was first introduced by Karl Friedrich Burdach in 1822.

The regions described as amygdala nuclei encompass several structures with distinct connectional and functional characteristics in humans and other animals. Among these nuclei are the basolateral complex, the cortical nucleus, the medial nucleus, the central nucleus, and the intercalated cell clusters. The basolateral complex can be further subdivided into the lateral, the basal, and the accessory basal nuclei. Anatomically, the amygdala, and more particularly its central and medial nuclei, have sometimes been classified as a part of the basal ganglia.

In one study, electrical stimulations of the right amygdala induced negative emotions, especially fear and sadness. In contrast, stimulation of the left amygdala was able to induce either pleasant (happiness) or unpleasant (fear, anxiety, sadness) emotions. Other evidence suggests that the left amygdala plays a role in the brain's reward system.

Each side holds a specific function in how we perceive and process emotion. The right and left portions of the amygdala have independent memory systems, but work together to store, encode, and interpret emotion.

The right hemisphere is associated with negative emotion, it plays a role in the expression of fear and in the processing of fear-inducing stimuli. Fear conditioning, which occurs when a neutral stimulus acquires aversive properties, occurs within the right hemisphere. When an individual is presented with a conditioned, aversive stimulus, it is processed within the right amygdala, producing an unpleasant or fearful response. This emotional response conditions the individual to avoid fear-inducing stimuli and more importantly, to assess threats in the environment.

The right hemisphere is also linked to declarative memory, which consists of facts and information from previously experienced events and must be consciously recalled. It also plays a significant role in the retention of episodic memory. Episodic memory consists of the autobiographical aspects of memory, permitting recall of emotional and sensory experience of an event. This type of memory does not require conscious recall. The right amygdala plays a role in the association of time and places with emotional properties.

The right amygdala is also linked with taking action as well as being linked to negative emotions, which may help explain why males tend to respond to emotionally stressful stimuli physically. The left amygdala allows for the recall of details, but it also results in more thought rather than action in response to emotionally stressful stimuli, which may explain the absence of physical response in women.

Joseph Le Doux [99] has shown that the amygdala creates direct connections between sensory inputs such specific sounds or pain responses during Pavlovian training. This suggests that when we receive fear responses specifically our reflexes can be programmed or reprogrammed.

4.1.8.1.2   Hippocampus

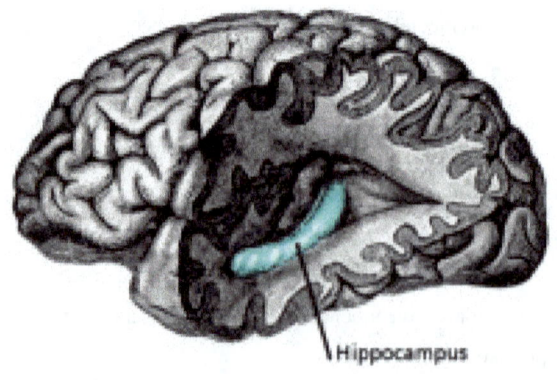

Figure 74 Hippocampus

The hippocampus (via Latin from Greek ἱππόκαμπος, "seahorse") is a major component of the brain of humans and other vertebrates. Humans and other mammals have two hippocampi, one in each side of the brain. The hippocampus is part of the limbic system, and plays important roles in the consolidation of information from short-term memory to long-term memory, and in spatial memory that enables navigation. The hippocampus is located under the cerebral cortex in the allocortex, and in primates it is in the medial temporal lobe. It contains two main interlocking parts: the hippocampus proper (also called Ammon's horn) and the dentate gyrus.

In Alzheimer's disease (and other forms of dementia), the hippocampus is one of the first regions of the brain to suffer damage; short-term memory loss and disorientation are included among the early symptoms. Damage to the hippocampus can also result from oxygen starvation (hypoxia), encephalitis, or medial temporal lobe epilepsy. People with extensive, bilateral hippocampal damage may experience anterograde amnesia: the inability to form and retain new memories.

Since different neuronal cell types are neatly organized into layers in the hippocampus, it has frequently been used as a model system for studying neurophysiology. The form of neural plasticity known as long-term potentiation (LTP) was initially discovered to occur in the hippocampus and has often been studied in this structure. LTP is widely believed to be one of the main neural mechanisms by which memories are stored in the brain.

In rodents as model organisms, the hippocampus has been studied extensively as part of a brain system responsible for spatial memory and navigation. Many neurons in the rat and mouse hippocampus respond as place cells: that is, they fire bursts of action potentials when the animal passes through a specific part of its environment. Hippocampal place cells interact extensively with head direction cells, whose activity acts as an inertial compass, and conjecturally with grid cells in the neighbouring entorhinal cortex.

4.1.8.1.3 Septum

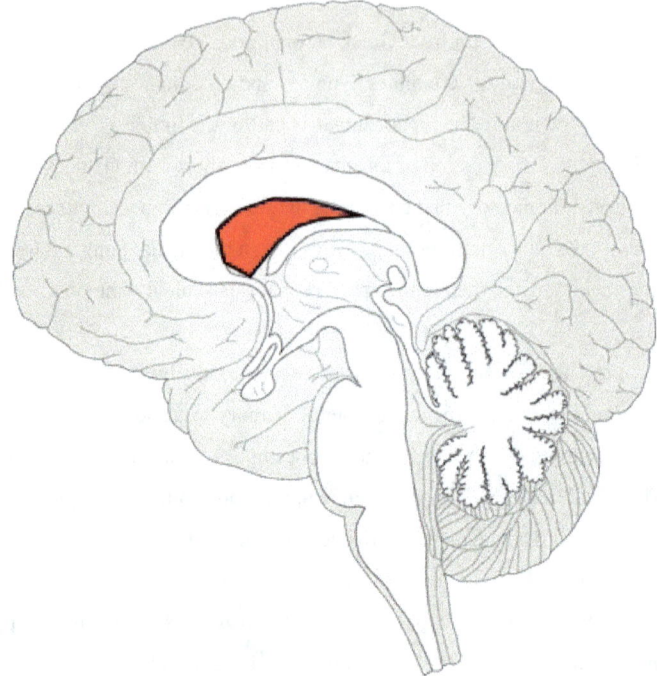

Figure 75 Septum

Septum Verum (true septum) is a region in the lower medial part of the telencephalon that separates the two cerebral hemispheres. The human septum consists of two parts: the septum pellucidum (translucent septum), a thin membrane consisting of white matter and glial cells that separate the lateral ventricles, and the lower, precommisural septum verum, which consists of nuclei and grey matter. The term is sometimes used synonymously with Area Septalis, to refer to the precommisural part of the lower base of the telencephalon. The Septum verum contains the septal nuclei. The septum is considered a part of the limbic system, mediating the connection between the cortex and subcortical limbic nuclei. The septum projects fibres to the hypothalamus, hippocampus, amygdala, reticular formation and olfactory cortical areas, suggesting a role in limbic regulation.

While the exact function remains controversial, the septum is considered a pleasure zone in animals, studies have shown that stimulation of the septal area can bring feelings of satisfaction to euphoria and damage can cause hyperactivity and fury.

4.1.8.2 Frontal Lobe

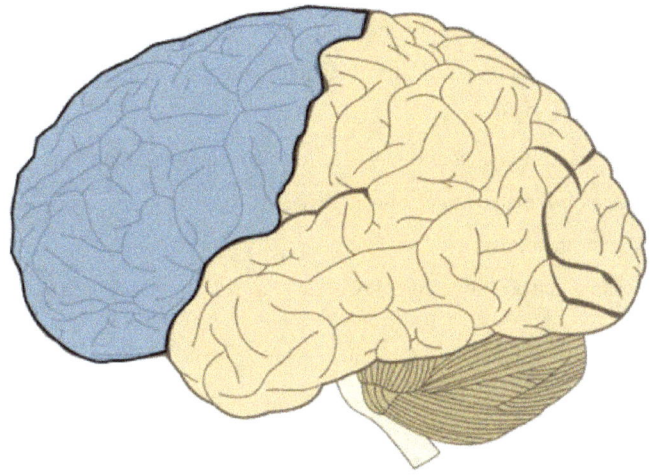

Figure 76 Frontal Lobe

The front section of the brain is the Frontal Lobe, in front of the parietal and temporal lobes. It is separated from the parietal lobe by a groove between tissues called the central sulcus and from the temporal lobe by a deeper groove called the lateral sulcus (Sylvian fissure). The frontal lobes can be divided into five zones: the primary motor, premotor, prefrontal, paralimbic, and limbic zones. These areas relate to reasoning, motor skills, higher level cognition, and expressive language. The motor cortex lies at the back of the frontal lobe, near the central sulcus, which controls voluntary movements of specific body parts. This area of the brain receives information from various lobes of the brain and utilizes this information to carry out body movements. Damage to the frontal lobe can lead to changes in sexual habits, socialization, and attention as well as increased risk-taking. This area of the brain does not mature until people are in their 20's, earlier for women and later for men, which explains some risk taking activities in later teenage years.

The frontal lobe contains most of the dopamine neurons in the cerebral cortex. The dopaminergic pathways are associated with reward, attention, short-term memory tasks, planning, and motivation. Dopamine tends to limit and select sensory information arriving from the thalamus to the forebrain.

4.1.8.2.1 Prefrontal cortex

The prefrontal cortex helps people set and achieve goals. It receives input from multiple regions of the brain to process information and adapts accordingly. The prefrontal cortex contributes to a wide variety of executive functions, including:

- Focusing one's attention
- Predicting the consequences of one's actions; anticipating events in the environment
- Impulse control; managing emotional reactions
- Planning for the future
- Coordinating and adjusting complex behaviours ("I can't do A until B happens")

4.1.8.3 Parietal Lobe

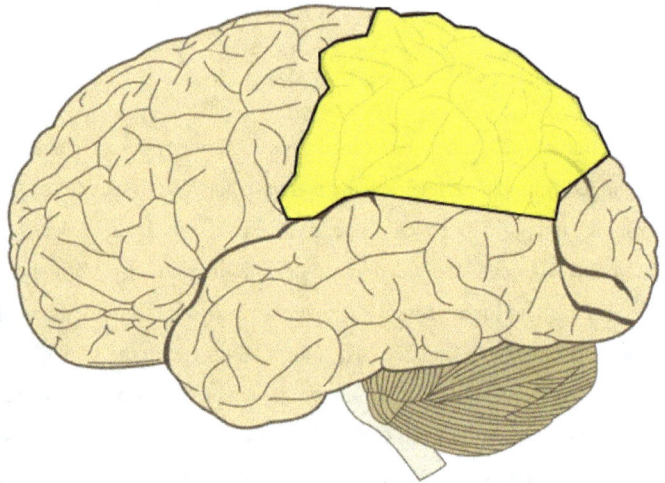

Figure 77 Parietal Lobe

The top middle section of the brain is the parietal lobe. The name comes from the parietal bone, which is named from the Latin paries-, meaning "wall". It is positioned above the temporal lobe and behind the frontal lobe and central sulcus. The parietal lobe integrates sensory information among various modalities, including spatial sense and navigation (proprioception), the main sensory receptive area for the sense of touch in the somatosensory cortex which is just posterior to the central sulcus in the postcentral gyrus, and the dorsal stream of the visual system. The major sensory inputs from the skin (touch, temperature, and pain receptors), relay through the thalamus to the parietal lobe.

Several areas of the parietal lobe are important in language processing. The somatosensory cortex can be illustrated as a distorted figure – the cortical homunculus (Latin: "little man") in which the body parts are rendered according to how much of the somatosensory cortex is devoted to them.

In terms of the design this is where a self-model would be most useful and would allow the sensory information to be mapped to locations on the body.

4.1.8.3.1 Primary Sensory Cortex

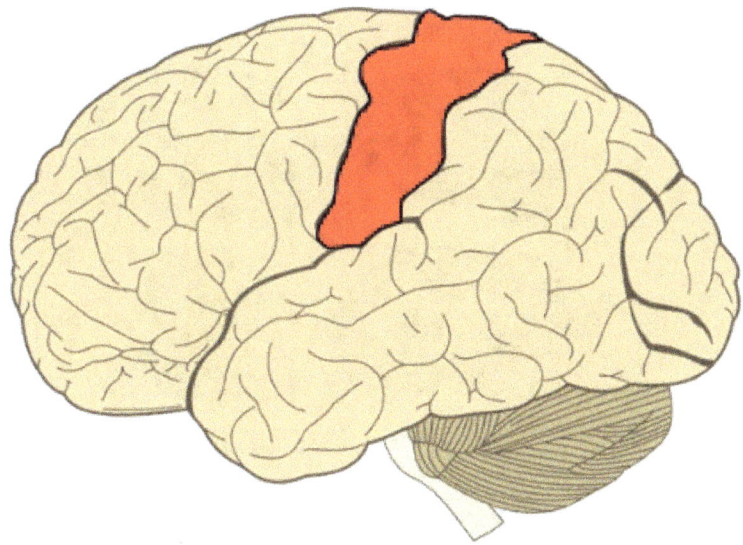

Figure 78 Primary Sensory cortex

The primary somatosensory cortex is located in the postcentral gyrus, and is part of the somatosensory system. It receives the bulk of the thalamocortical projections from the sensory input fields. At the primary somatosensory cortex, tactile representation is orderly arranged (in an inverted fashion) from the toe (at the top of the cerebral hemisphere) to mouth (at the bottom). Each cerebral hemisphere of the primary somatosensory cortex only contains a tactile representation of the opposite (contralateral) side of the body. The amount of primary somatosensory cortex devoted to a body part is not proportional to the absolute size of the body surface, but, instead, to the relative density of cutaneous tactile receptors on that body part. The density of cutaneous tactile receptors on a body part is generally indicative of the degree of sensitivity of tactile stimulation experienced at said body part. For this reason, the human lips and hands have a larger representation than other body parts.

Hypothesis:

The primary sensory cortex, represents an input map of the body. This may be represent part of a self-model. The self-model allows sensory information to be mapped onto a model of the body so that it is possible to interpret incoming signals with respect to each other with and motor outputs. Close by is the Primary motor cortex that controls outputs to the body. I suspect that these two parts of the brain for an Input output model of the body and form the central component of the self-model. The self-model can be used to predict the behaviour of the real world body, allowing strategic planning and learning to occur.

4.1.8.3.2   Secondary Sensory Cortex

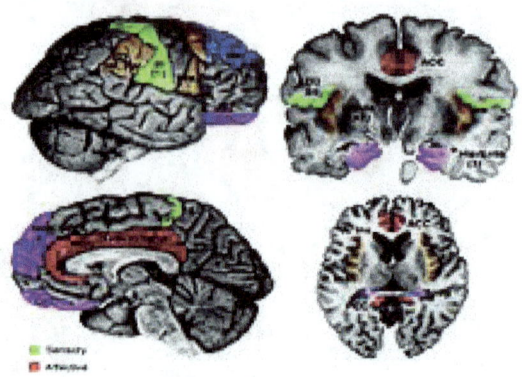

Figure 79 Secondary Sensory Cortex

The human secondary somatosensory cortex is a region of cortex in the parietal operculum on the ceiling of the lateral sulcus. It is believed to perform higher order functions including sensorimotor integration, integration of information from the two body halves, attention, learning and memory. Experiments involving ablation of the second somatosensory cortex in primates indicate that this cortical area is involved in remembering the differences between tactile shapes and textures. Functional neuroimaging studies have found S2 activation in response to light touch, pain, visceral sensation, and tactile attention.

The somatosensory association cortex is directly posterior to the sensory cortex in the superior parietal lobes. This receives synthesized connections from the primary and secondary sensory cortices. These neurons respond to several types of inputs and are involved in complex associations. Damage can affect the ability to recognize objects even though the objects can be felt (tactile agnosia). The Secondary somatosensory cortex (SII) has been recently implicated in processes that require high-level information integration, such as self-consciousness, social relations, whole body representation, and metaphorical extrapolations. [100]

Hypothesis:

{Evil Genius: Haha!! I was right!!!}

This research [100] does indicate that there is a self-based model, higher level cognition and potentially abstract relationships in the brain. It should not be much of a leap because we do have these capabilities, hence there must exist some structure of the brain that performs these tasks. The main questions is not if these structures exist but where, and is it a single structure or multiple structures that interact.

4.1.8.3.3 Primary Motor Cortex

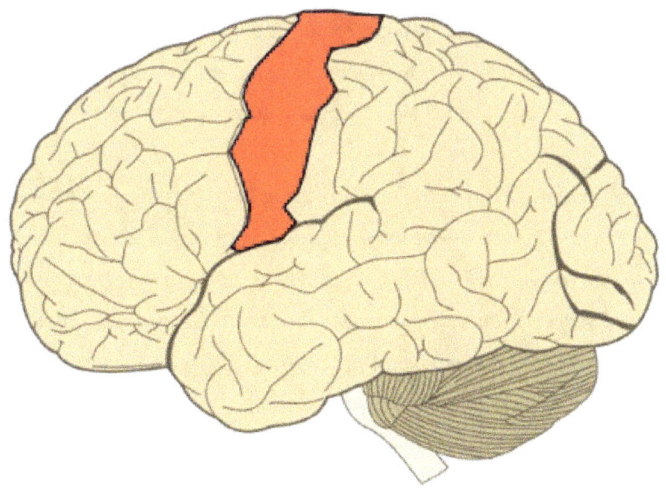

Figure 80 Primary Motor Cortex

The primary motor cortex is a brain region that in humans is located in the dorsal portion of the frontal lobe. The upper motor neurons in the cerebral cortex reside in several adjacent and highly interconnected areas in the frontal lobe, which together mediate the planning and initiation of complex temporal sequences of voluntary movements. It is the primary region of the motor system and works in association with other motor areas including premotor cortex, the supplementary motor area, posterior parietal cortex, and several subcortical brain regions, to plan and execute movements. These cortical areas all receive regulatory input from the basal ganglia and cerebellum via relays in the ventrolateral thalamus as well as inputs from the somatic sensory regions of the parietal lobe. Although the phrase "motor cortex" is sometimes used to refer to these frontal areas collectively, more commonly it is restricted to the primary motor cortex, which is located in the precentral gyrus. Primary motor cortex is defined anatomically as the region of cortex that contains large neurons known as Betz cells. Betz cells, along with other cortical neurons, send long axons down the spinal cord to synapse onto the interneuron circuitry of the spinal cord and also directly onto the alpha motor neurons in the spinal cord which connect to the muscles.

At the primary motor cortex, motor representation is orderly arranged (in an inverted fashion) from the toe (at the top of the cerebral hemisphere) to mouth (at the bottom) along a fold in the cortex called the central sulcus. In many ways similar to the Primary Sensory Cortex which maps to inputs as opposed to the outputs of the Primary Motor Cortex. The amount of primary motor cortex devoted to a body part is not proportional to the absolute size of the body surface, but, instead, to the relative density of cutaneous motor receptors on said body part. The density of cutaneous motor receptors on the body part is generally indicative of the necessary degree of precision of movement required at that body part. For this reason, the human hands and face have a much larger representation than the legs.

Hypothesis:

The primary motor cortex, represents an output map of the body. This may be represent part of a self-model. The self-model allows sensory information to be mapped onto a model of the body so that it is possible to interpret incoming signals with respect to each other and motor outputs. I suspect that these two parts of the brain for an Input output model of the body and form the central component of the self-model. The self-model can be used to predict the behaviour of the real world body, allowing strategic planning and learning to occur.

4.1.8.4   Occipital Lobe

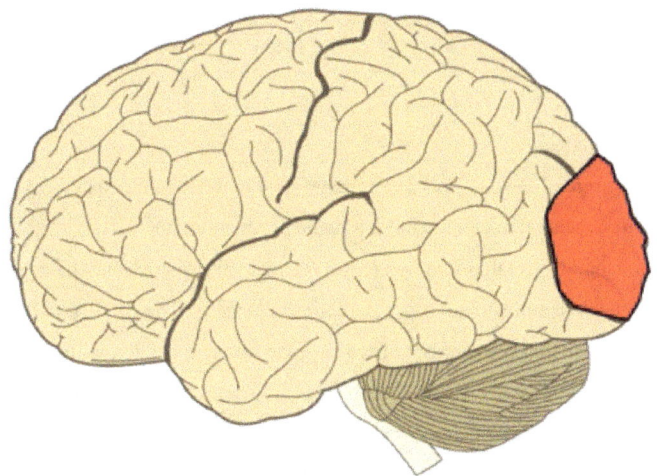

Figure 81 Occipital Lobe

The back portion of the brain is the occipital lobe it is the smallest of the four lobes of the cerebral hemisphere. The primary visual cortex is located in the occipital lobe. It receives and interprets information from the retinas of the eyes. This is where stereo image matching would be necessary allowing depth perception. There are many extra-striate regions, and these are specialized for different visual tasks, such as visuospatial processing, colour differentiation, and motion perception. Damage to this lobe can cause visual problems such as difficulty recognizing objects, an inability to identify colours, and trouble recognizing words.

Hypothesis:

The visual processing that occurs in the occipital lobe performs object detection and recognition. Object recognition would require access to symbolic "data" for visual data to be matched against. This would explain why recognition is disrupted if the occipital lobe is damaged.

4.1.9   Language Production and Comprehension

Research covered by Fernyhough [22] suggests that disruption to Broca's or Wernicke's areas (or associated areas) may cause schizophrenic symptoms including auditory hallucinations. In other words we might experience auditory hallucinations as Wernicke's area produces an internal voice that is not recognised by Broca's area as a voice that is self-generated.

Hypothesis:

Speech is produced internally from Broca's area and we interpret the same speech in Wernicke's area to generate or reinforce the ideas we are attempting to express in what I refer to as "the consciousness loop".

4.1.9.1   Broca's Area

The precise role of Broca's area in language production is still debated. Evidence suggests that damage to Broca's area can disrupt language production, but nobody is quite sure exactly what specific language-related function is lost to cause that disruption. Different regions of the Broca's area specialize in various aspects of comprehension. The anterior portion helps with semantics, or the meaning of words, while the posterior is associated with phonology or how words sound. Some have asserted Broca's area is involved with producing motor movements that allow speech to be produced. Broca's area is also necessary for language repetition, gesture production, sentence grammar, and fluidity and is involved in interpreting others' actions.

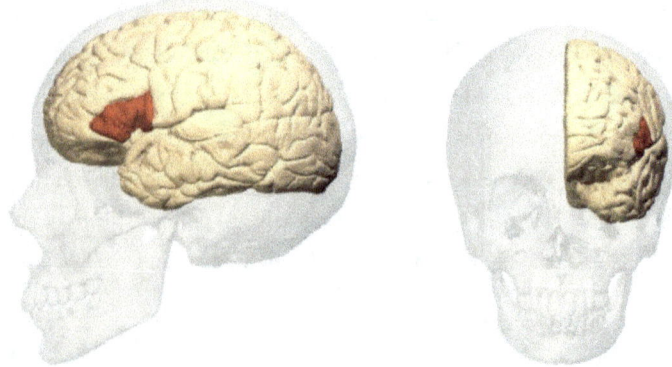

Figure 82 Broca's Area

4.1.9.2   Wernicke's Area

The term "Wernicke's area" is most often used as an anatomical label for the gyri forming the lower posterior left sylvian fissure. Thus, it lies close to the auditory cortex. Although traditionally this region was held to support language comprehension, modern imaging and neuropsychological studies converge on the conclusion that this region plays a much larger role in speech production. Because the Wernicke area is responsible for the comprehension of written and spoken language, damage to this area results in a fluent but receptive aphasia. Receptive aphasia may be best described as one who is unable to comprehend/express written or spoken language. The patient will most commonly have fluent speech, but their words will lack meaning.

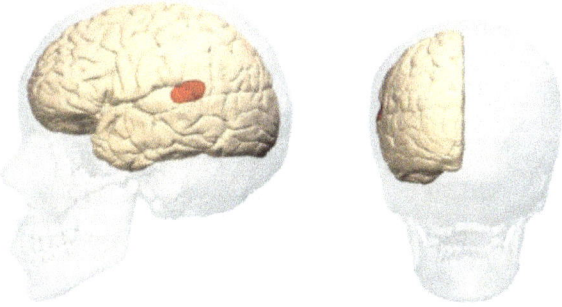

Figure 83 Wernicke's Area

4.1.10   The Default Mode Network

The default mode network (DMN), (default network or default state network), is a brain network that include the medial prefrontal cortex, posterior cingulate cortex, inferior parietal lobule, angular gyrus, the lateral temporal cortex, hippocampal formation, and the precuneus. [101]–[104]

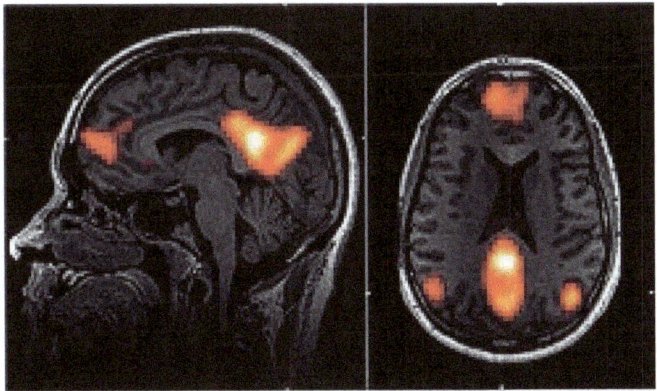

Figure 84 The Default Mode Network

The default mode network (DMN) consistently decreases its activity when compared with activity during these relaxed "non-task" states. Its discovery was an unexpected consequence of brain-imaging studies first performed with positron emission tomography in which various novel, attention-demanding, and non-self-referential tasks were compared with quiet repose either with eyes closed or with simple visual fixation. The discovery of the default mode network reignited a longstanding interest in the significance of the brain's ongoing or intrinsic activity. Presently, studies of the brain's intrinsic activity, popularly referred to as resting-state studies, have come to play a major role in studies of the human brain in health and disease. The brain's default mode network plays a central role in this work.

It is best known for being active when a person is not focused on the outside world and the brain is at wakeful rest, such as during daydreaming and mind-wandering. It can also be active during detailed thoughts related to external task performance. Other times that the DMN is active include when the individual is thinking about others, thinking about themselves, remembering the past, and planning for the future.

Though the DMN was originally noticed to be deactivated in certain goal-oriented tasks and is sometimes referred to as the task-negative network, it can be active in other goal-oriented tasks such as social working memory or autobiographical tasks. The DMN has been shown to be negatively correlated with other networks in the brain such as attention networks.

4.2 The Brain Vs The Design

Now we have two models of we can compare. The first is from the biological models (especially from Baars and Edelman). This explains where some of the functions we know of occur in the brain. The second model is "The design", which is more of a functional model of the processing required to create something resembling consciousness. The ideas here is that we can compare the two models and see what kind of similarities exist.

Take this with a grain of salt as they say. It is very much guess work! That said lining up the components of design with components of brain where they have similar functionality, from reading my hypotheses in the previous sections, allows us to see if the design is actually a sensible approach. This also gives people a framework to examine changes or additions to either model and allows the changes to be compared and tested. So while I am presenting this as fact this is the best guess I can make based on my *{limited}* understanding of the operation of the brain. From the structure of the brain the first thing we should notice is that there is not a one to one relationship between the structure of the brain and the operations of "The Design". This is not hugely surprising as the brains functions are often quite distributed and hence difficult to pin down with the research that has been done.

4.2.1 Matching between Biological brains and the design

Baars theory of the global workspace, describes the ERTAS (Extended Reticular Thalamic Activating System) as the location of the brain where consciousness occurs. This is definitely close to the location of sensory input and qualia matching. The Thalamus provides a connection between the Cerebral cortex and the brainstem it helps regulate sleep, alertness and wakefulness it may be related to consciousness in general. The assumption we make from the design is that two processes need to occur to create consciousness: Internal Qualia generation and Qualia matching/analysis. The Thalamus is in a sensible location for this, but I am not sure that it has all of the functions necessary. Specifically the thalamus does not seem to be related to analysis of sensory data or simulating actions in the world.

While the Thalamus, Pons and Reticular formation have some association with consciousness, this may be related to "sleep cycle" than what we are describing in the design. The sleep cycle is more related to the day night cycle or the management of waking and sleeping activities. So while this could potentially shut down consciousness so that a creature could sleep, as opposed to what we are discussing in the design of the sensory generation that provides an "internal window of the world".

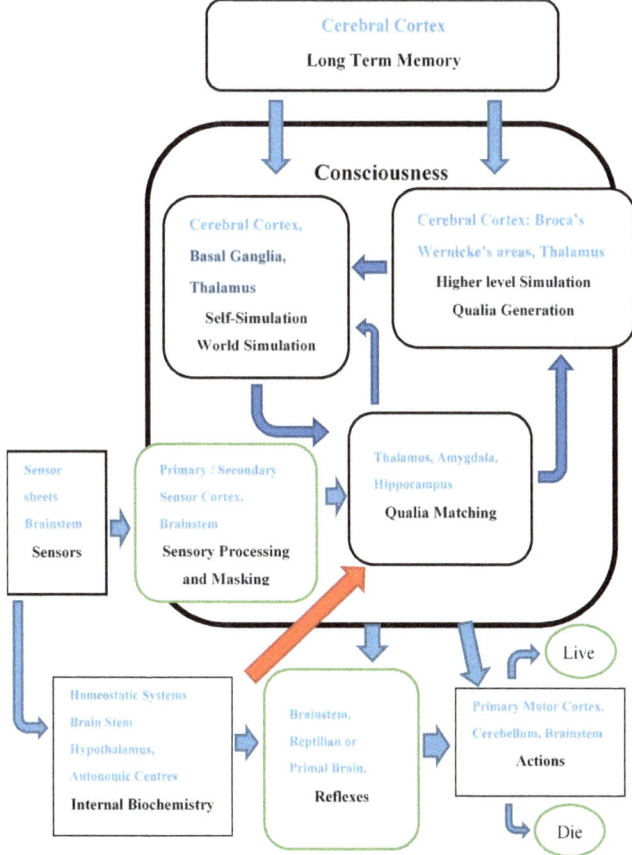

Figure 85 Design of Consciousness mapped to Brain areas

The most basic functions of the brain seem to match up with processing described in the design quite well, including:

- Sensory processing operating in the sensor sheets of the brainstem, including the pons and medulla.
- Reflexes seem to operate from the reptilian brain and possibly the brainstem
- Output actions run from the Primary motor cortex, the cerebellum output through the Brainstem.

Other parts of the consciousness puzzle might include the Basal Ganglia, Midbrain, Cerebellum and Hippocampus. Some research shows that the Basal Ganglia is responsible for providing updates to the thalamus at least for tracking movements. The midbrain is again close to the thalamus and reticular formation, and is along the pathway to the transmission of signals to the thalamus. The cerebellum acts as a control system for balance and smoothing body movements. The same algorithms may also be used for predicting body movements or other objects. The Hippocampus is central to working memory and potentially ordering or accessing long term memory [105].

There is some discussion on the location of working memory. Some suggest specific areas including the ventral and dorsal lateral prefrontal cortex [106], superior occipital gyrus [107], while others show that it can be more distributed [108], [109]. Part of the difficulty with working memory in terms of brain scans is that we cannot be sure that parts of the brain being activated are actually references to long term memory that are being modified in some other area of the brain which operates as working memory. This could also operate in conjunction with masking operations that seem to exist in the Pons and medulla. Thus far I have not found any agreement on which organ this might be if indeed it is a single organ. So short version: Dunno.

Sensory inputs most likely travel from the body through the Pons, Medulla through the Reticular Formation, midbrain and into the Thalamus. Sensor processing, filtering and masking likely occurs along this pathway at least for signals from the body. The other major sensors we have are the auditory and visual senses. Visual senses are connected closely to the occipital lobe in the brain. My guess is that this has as much to do with processing stereo vision as object recognition. The parietal and occipital lobes deal primarily with sensory inputs so they probably relate most to sensory processing and possibly qualia matching and interpretation. The theory is that these parts of the brain would both receive sensory data (qualia from the real world) and qualia generated from simulations. These two datasets would be masked and matched in their associated lobes, then processed by the higher level symbolic conceptual and abstract processors, which would be distributed into other lobes of the brain. The Frontal lobes seem to be more related to higher order abstract processing. Again, this may be as much knowledge as active working memory, or working memory may operate in the thalamus or distributed anywhere along the sensory processing path.

The Thalamus, Amygdala and Hippocampus are my best guess for the location of Qualia Matching. Most likely of these is the Thalamus which is central to incoming sensory data and long term memory. There is also some evidence that these areas are related to different variations of working memory, including motor and associative activities [110].

The processes of Self Simulation, World Simulation and Qualia Generation seem to be closely related to the Cerebral Cortex, Basal Gangliaand Thalamus, Parr and Friston's work [110] also shows signal loops between cortical structures and the Thalamus, via the Basal ganglia. Again not something I expected to find, but it seems consistent with the design hypothesis. My guess is that the cerebral cortex may be storing information or even what we might call "algorithms" or "update algorithms" that are used in the basal ganglia or thalamus. There is a lot of research showing that the different lobes of the cerebral cortex correspond to different types of memories.

Higher level Simulation including abstract processing seem to relate to the Frontal lobes and we know that much of language processing occurs in Broca's and Wernicke's areas.

From the overview of medical research on brain function it seems comparable with the operation of the design of consciousness, at least at a surface level.

Parr and Friston in "Working memory, attention, and salience in active inference" [110] present several diagrams of where brain activity occurs during working memory tasks. It shows that signals trace from the cerebral cortex to the thalamus and back in a loop. So while the hypothesis of the consciousness loop might be a bit out there, it does seem to have some supporting evidence.

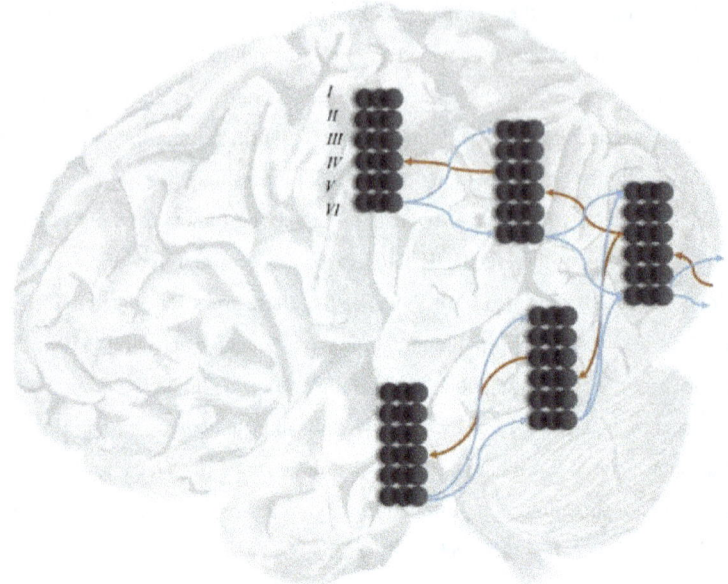

Figure 86 Cortical hierarchies [110]

http://creativecommons.org/licenses/by/4.0/

"There is a systematic, stereotyped pattern of laminar-specific connectivity in the cerebral cortex. Layer IV receives ascending (extrinsic) input, and has (intrinsic) connections to more superficial layers (II/III) in the same region. This layer provides (extrinsic) projections to layer IV in higher cortical areas and (intrinsic) connections to deep layer VI. Layer VI projects to both deep and superficial layers in lower regions, and this provides a pattern of descending (extrinsic) connections. These patterns of (extrinsic) connectivity can be used to map out hierarchies in the brain. The hierarchies in the two streams shown here ascend from the right of the image to the left. "[110]

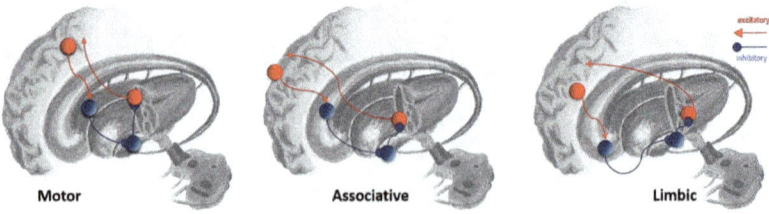

Figure 87 Parallel cortico-subcortical loops [110]

http://creativecommons.org/licenses/by/4.0/

"Cortical areas target particular regions of the basal ganglia. This cortical input is the start of a loop through the basal ganglia, back to the same cortical area via the thalamus. For simplicity, only the direct pathway through the basal ganglia is shown in these figures. Connections from the dopaminergic midbrain are not shown, but these target the caudate and putamen." [110]

There are a few features to note here. The main one is the loop from various places in the Cerebral cortex, depending on the task at hand, to the Thalamus and back. The other is the pathway along the Basal ganglia. The guess is that we access parts of our cortex as part of working memory and use a simulation to make predictions about what will happen to the objects in working memory. We noted in Chapter 4.1.6 that the basal ganglia was associated with planning, movement and memory, which is consistent with this theory, but also means that it could be used for interpolation or extrapolations that are not directly related to movement. Again I take this a reasonable evidence for the design.

4.2.2   Conscious ignition

Dehaene [94] has shown that just before we become conscious of a particular sensation we see a cascading activation of neurons. This signal referred to as the P3 wave starts with sensory related areas of the brain and moves towards the frontal lobe which are related to more abstract reasoning and planning. The P3 wave seems to relate to repressing signals from activated neuron groups.

Bidirectional nature of Neurons allows neuron groups to "reproject" their activation back down the same channels. This would allow the selected groups to rebuild the associated detections from lower levels into some other area of the brain using the same wiring.

Dehane also describes the P3 wave as an inhibitory signal repressing the priming of neuron clusters activated by sensory analysis. My guess is that the majority of Primed clusters are repressed, but those related to concepts activated in short term memory are amplified. As there would be much fewer concepts activated by short term memory (five plus or minus two) versus any irrelevant activations (Thousands, maybe millions) the amplification signal would be much harder to detect especially compared to the P3 wave.

We still don't know "where" consciousness occurs. Is it purely created from the signals inside the cerebral cortex? Does the "simulation" of the world exist in the cortex, the thalamus or is it distributed between all of the parts of the brain and we only feel conscious when all of the components interact? At this point I still don't think we know for certain, but we are definitely getting closer to an answer.

4.2.3 Conceptual activation

The same neuron groups can be activated by different stimulus. Visual and audio senses can activate the same neuron group. This suggests that groups of neurons could represent concepts.

Visual processing also follows a similar pattern. Lower level processes detect specific features, as we look further into the processing, we detect groups of features, faces and eventually specific people. The same with auditory processing, sounds then syllables, then words, then sentences.

These sensory channels can then combine when the concepts we are considering have both auditory and visual features. We can activate the concepts of a piano by seeing a piano, seeing the word "piano", by hearing a piano playing or hearing the word piano. These would all activate the same neuron cluster that could then be reproduced back into the sensory stages allowing us to "imagine" a piano when any of the sensory triggers occur.

4.3 Comparing Biology and the Design

What I have been attempting to do in this section is cover many aspects of biological systems that are related to consciousness and demonstrate that either these phenomena can be shown to exist in an artificial system or, more preferably, show how this phenomena could be present in the design.

These phenomena include

- Sleep
- Dreaming
- Memory
- Learning

There has been a lot of research on the relationship between consciousness, memory, learning and sleep since the 1990's. [111], [112]. As such it is fairly well established that consciousness allows new things to be learned. Sleep also plays a significant role especially in storing long term memories.

4.3.1 Anaesthetics

We know anaesthetics can be used to "shut down" human consciousness and allow people to safely recover, similar to how sleep works. Anaesthetics are used during surgery to help make sure people are not conscious during what would be extremely traumatic experience. There are several different types of anaesthetics that operate on different areas or functions of the brain. The notable common effect of these drugs is that they all suppress consciousness. With reference to the design we can look for potential processes to target that would create a similar effect to an anaesthetic.

From Figure 88 we can see that 4 processes are involved. If we suppressed the external sensory qualia generation the simulation loop can still be active. This is more where pain killer would be active as the actually suppress the inputs to the system. Notably this would not remove consciousness.

If we suppress Qualia interpretation the Simulation and Sensory qualia could be ignored and the output to the simulation would be shut down. This may result in the simulation being active but unable to perceive anything. This could result in the simulation being inactive or it could result in a "sleep-like" state. There have been many reports from people who have undergone surgery with anaesthesia and felt that they woke up during the surgery, but been unable to move. Suppressing Qualia interpretation may cause a similar experience in this design.

When Curare was originally discovered doctors and animal researchers attempted to use it for surgery [113], [114], the story is that to the doctors and their patient's horror they discovered that the patient was fully conscious during surgery, but unable to move. As such, it was banned from use with animal experimentation. Interestingly Curare is, as far as I can tell, still allowed in surgery as it is extremely good at reducing muscle actions during surgery. Thankfully, it is used with other anaesthetics to ensure the patient does not feel any cutting.

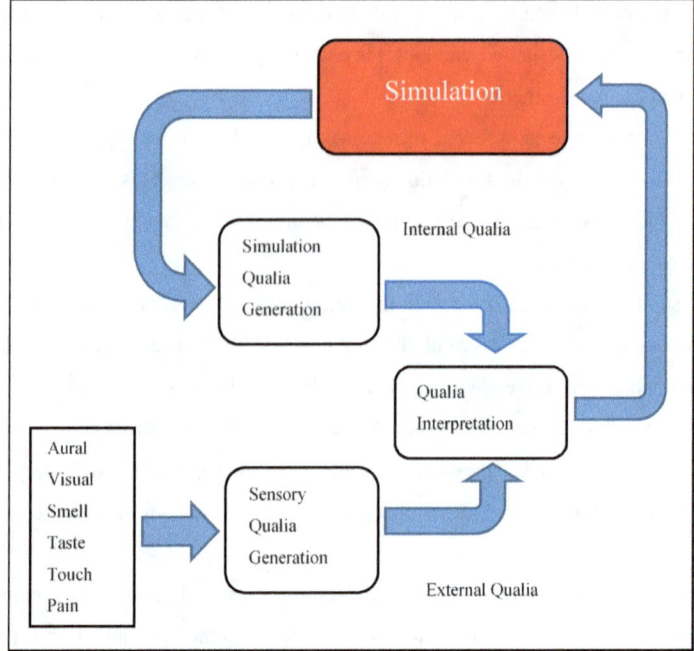

Figure 88 Detail of Consciousness loop Highlighting Anaesthetics

That can also be mapped onto the model to see what is affected. Ideally, just reflex muscle actions are suppressed, but in surgery not the autonomic functions. So most like it represses nerve signals in the brain stem and possibly cerebellum.

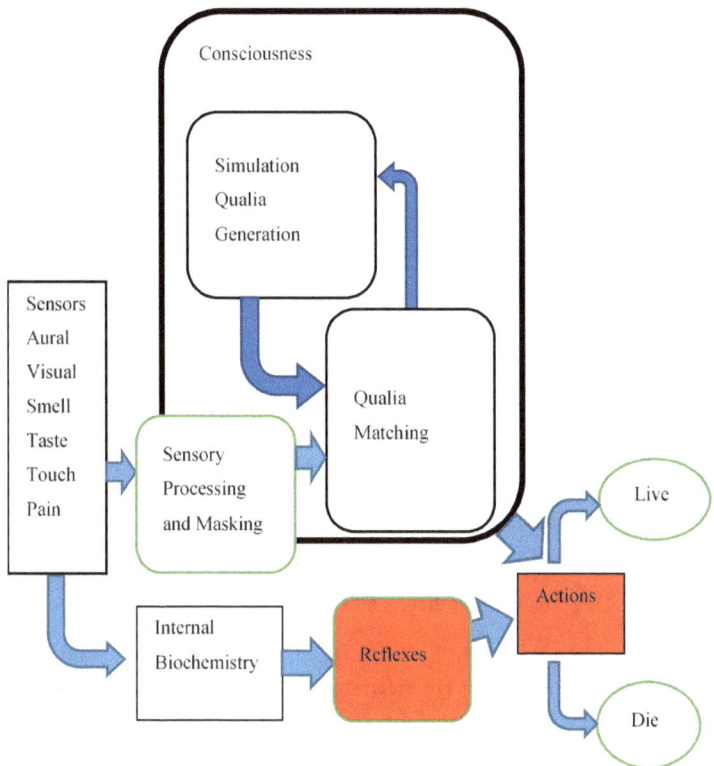

Figure 89 Highlighting the effect of Paralytics

If we suppress the Simulation itself we stop interpreting the world around us. This is probably the most likely target for anaesthetics as it would completely shut down consciousness.

If we suppress Simulation Qualia generation it is likely that we would still feel the external stimulus, but not realise that the stimulus is effecting the "body" of the associated consciousness.

Each one of these targets might not be enough by itself, however both areas might need to be targeted. From the model these functions match those from the Thalamus. The Thalamus is also shown to be affected by anaesthetics and has led to the theory that the Thalamus is central to consciousness [115]–[118].

4.3.2 Sleep

We know that the brain either is a neural network or can be represented by a neural network. The interesting thing about neural networks in the study of artificial intelligence is that they can 'learn'. The neural network can be reprogrammed, but this is actually a complex process using training data and testing cycles.

We have long known that sleeping is helpful for learning in animals and especially humans. What is interesting is that it is possible the same states exist in biological brains.

James Pardey et. al. "A new approach to the analysis of the human sleep/wakefulness continuum" [119] examined EEG readings during sleep. The various stages of sleep were classified using a self-organizing feature map.

These stages were further grouped into three main separate states: Awake, Sleep stages 1-4 and REM sleep. It is possible that these states correspond to the states to reprogram a neural network.

The 'Awake' state is obvious. The sleep stages 1 to 4 occur as people fall asleep and the body is placed into an inactive state so that it does not react to mental stimulus. The REM state, or Rapid Eye movement. During this state as the name suggests they eyes move rapidly.

To reprogram a neural network the network needs to be placed in a 'shutdown state' where inputs can be applied and the outputs of the neural network can be ''back propagated' into the network nodes. This procedure allows the connections between nodes to be adjusted.

The next phase of programming is to test the network to determine if the network is operating correctly. To do this sample data inputs are given to the network and the output of the network compared to the expected outputs.

Once the neural network is programmed and tested it can be used in 'real' situations with active input data. The point is that for the AI system to learn and adapt to new situations it needs to be shut down and reprogrammed using specific procedures.

Comparing the states for sleep and the states of operating a neural network show some similarities. This should not be too surprising as the artificial intelligence system is based on the biological system.

The awake state of a biological brain and the operational state of an AI are obviously equivalent. The two other states are slightly more difficult to classify as we don't know exactly what is occurring in a human brain. I think it is most likely that the REM state in biological represents the reprogramming state in AI. This leaves the sleep stages 1-4 as the testing state. It is possible that the REM could be testing and sleep 1-4 could be updating in AI. If we could determine that the REM stage was associated with particular hormones or learning.

This could mean that the rapid eye movement could represent an active process of updating or a rapid series of tests which may indicate that the brain is attempting to identify visual cues from an 'internal' simulation. The visual cues could be reconstructed from memories, but we may be able to take this further and suggest that the visual cues could be internally generated as those we experience in dreams. Dreams have been suggested to occur in either the sleep 1-4 or REM states. Again either way dreams are similar to what people experience while awake, but they do not often exactly represent what we remember from being awake, hence our brains must be able to create these experiences. This gets us back to the idea of the design consciousness.

The proposed design suggests that we humans and many other animals have an internal simulation of the real world including a self-model. This examination of sleep and dreaming shows that we have at least the potential to create new experiences in dreams that allow the brain to be tested and potentially updated.

So we have:

1. A biological system that we know can be conscious.

2. A few properties of a consciousness biological system that are similar to properties of neural networks.

This indicates that artificial systems can have the same functionality and properties as a biological system. This is not conclusive, but it does show that it is possible for an artificial system to have the same properties as a biological system.

In terms of consciousness we might end up with a neural network (deep learning) system that could be conscious, however this does not mean it will give us any deeper insight into what consciousness is. Even if we create an artificial system based on neural networks that seems to be conscious there will be no way of determining if it is as consciousness is an epistemic phenomena, it can only be understood by experiencing it. Demonstrating consciousness means we must be able to understand how it works, not just copy the behaviour.

4.3.3 *Dreaming*

Dreaming is another interesting behaviour of biological systems that most people have experienced. During a dream people often experience visual, aural, tactile and other experiences. These must be generated inside the brain somehow.

The design uses a simulation system that can generate "qualia" that could be experienced as such visual, aural and tactile experiences, essentially using the same connections as the real world senses via a feedback loop.

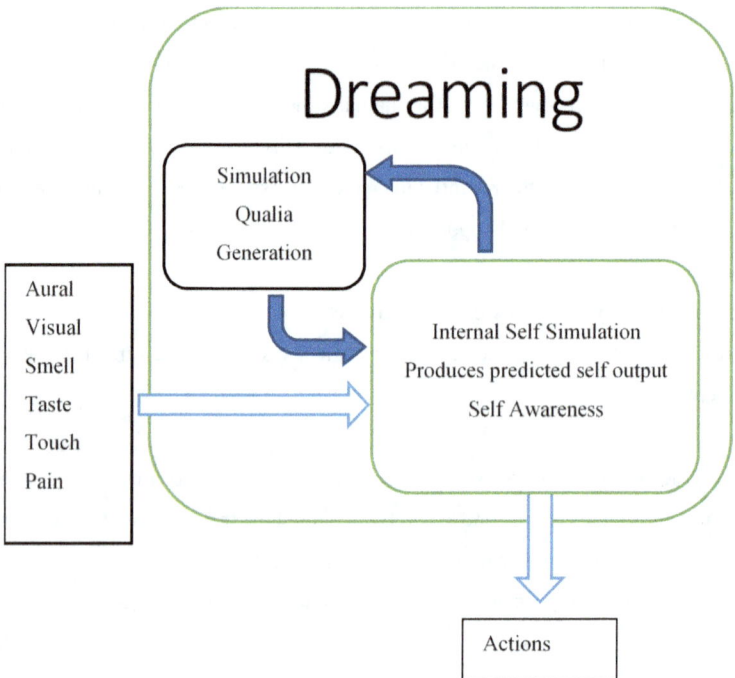

Figure 90 Dreaming system design

We know dreaming in a biological system occurs, but current research does not seem to show much about what is happening. This is often due to the fact that we don't know what data/information is being sent around the brain. We can only tell which areas of the brain are active under certain circumstances. The difficulty with dreaming is that people cannot report what they are thinking about as dreaming occurs.

This is where having a functional model comes in handy. So if we look at the design we can map onto this what is happening during dreaming. We know that external stimulus is suppressed and actions are suppressed. Part of the matching process may be suppressed as well as we want to ignore the incoming stimulus and suppress actions including reflex actions or low level actions that can be triggered by recognition.

We also know that when "lucid" dreaming people can remember objects, people, places, and other stimulus they recognise and stimulus that might not be recognised in the real world. So we know that these images can be created from "memories". The idea is that these are added into the simulation and run.

We also know that dreams can often be somewhat chaotic and do not necessarily follow an expected real world scenario. We can even get to nightmares where fear centres can be activated from the imagery produced. My suspicion with the chaos is similar to how we access memories via our "stream of consciousness" as discussed in Section Stream of Consciousness. The idea being we use common features to access memories. When we are dreaming we access memories to build up the dream "simulation" and figure out outcomes. These outcomes can also trigger memories as they will access related memories with similar accessors. As mentioned in the section on sleep, my guess is that dream are used to test out or possibly even adjust the neural network links in a back propagation system.

So while this may or may not be accurate it does show the utility of the functional model. If new differences are discovered they can be integrated into the existing model.

4.3.4   Vision

While vision itself is not a requirement of consciousness, there are a few testes that can be examined that relate to how our consciousness operates.

One of the weirdest things about vision is that you can have something in your visual field and not be aware of it. This is actually called "Inattentional Blindness" and was discovered by Arien Mack and Irvin Rock [120] and is the title of their book.

This research is actually one reason the design has several components. The idea is you are not "aware" of an object until three (3) processes have been performed.

- it has been recognised by your brain from incoming stimulus
- simulated, and
- matched to the incoming stimulus

This means there are 3 possible failure points for awareness. Recognising a new and unexpected stimulus, simulating the new stimulus and matching the simulation to the new stimulus.

My suspicion is that the simulation itself is the primary cause as we have limited working memory. As previously discussed we can usually only remember a small number of independent objects, typically 5 +/- 2. It also follows that we can only simulate a small number due to the complexity of potential interactions. The more objects and interactions potentially creates a factorial complexity which requires a significant amount of processing to calculate all possibilities.

So back to inattention, the more objects we need to simulate the harder it is to keep track of all of them. So if we are focused on a task and there is an unexpected stimulus it may be ignored by the simulation and masked out, especially if it does not interfere with the task.

In the other direction we have something called "blind-sight" which is discussed by many of philosophers and medical doctors in the consciousness field. In blind-sight the subjects claim not to be able to see. Yet in experiments they are shown to have an above guess (50/50) accuracy at guessing what object is in their visual field.

This suggests that the visual stimulus does at least get to the recognition stage, but is not populated into the simulation. This would allow the recognition stage to recognise the object, but the simulation will not reproduce the object and it will not be matched to incoming stimulus.

Another phenomena many consciousness researchers consider is conflicting stimulus. This occurs when different images are sent to our two eyes. Where conflicting stimulus occurs in our vision system, one set of inputs is typically ignored. As humans have developed stereo vision we have brain structures which deliberately align the eyes and match imagery between the eyes so that we can interpret the world around us in 3 dimensions. To resolve this conflict would require a different brain structure or a lot of processing. Creatures like chameleons who have independent eyes may have other methods of dealing with this issue. However we humans are not so lucky.

It makes sense that our ancestors often had to make decisions quickly. Which means that the best strategy would be to ignore one set of the conflicting inputs. With two eyes what occurs is that the images coming from one eye is ignored, people even become blind in the one eye even if the eye is fully functional but misaligned. Another example of this is if one of our ancestors received two sets of conflicting stimulus, one that said an area was safe and one that suggested a predator, it would be more sensible to assume a predator existed. As if they were wrong they would end up being eaten. The point being when you are dealing with a predator time matters, and making a quick decision is a sensible strategy. Hence with some redundancy in vision, i.e. we have two eyes, if there is conflicting data simply ignore one set.

My guess is that the same reflex occurs with abstract concepts. So if we have one set of ideas we are comfortable with and we are faced with another set that conflicts with them we would tend to reject any ideas we are uncomfortable with. What it would take is to consciously think through the abstract concepts to be able to integrate the less comfortable ideas even if they make logical or functional sense.

This might be too much of an extrapolation, but it might be possible. Take the idea with a grain of salt as they say.

4.3.5 *Reflexes*

Reflexes in life forms can occur in a few places. If you stab yourself with a needle *{please don't it hurts!}* or the doctor does the traditional tap on your knee with a small hammer, your muscles automatically contract. This action is called the reflex arc or unconditioned reflexes and has been shown to act inside the spinal cord. Other reflexes are found in the midbrain, medulla oblongata, such as coughing and sneezing.

Also as discussed in section 4.1.8.1.1 about the amygdala, was Joseph LeDoux [99] who found that Pavlovian fear responses caused the development of neural connections through the amygdala. These are referred to as conditioned reflexes and still occur automatically.

Subcortical structures are a group of diverse neural formations deep within the brain which include the diencephalon, pituitary gland, limbic structures and the basal ganglia. Some are related to emotion, which I argue are another type of reflex as they are not under conscious control. Yes there is some difference between the term I use here and the clinical definition of a reflex.

Even a learned behaviour such as catching a ball can be considered a reflex. When we attempt to catch a ball be do not actively calculate the trajectory of the ball and determine an inverse kinematic solution to catch the ball. You simply "decide" to catch the ball and move your hands to the right position, i.e. perform an action. This is sometimes referred to as a Conscious flow state.

When I trained in Ju-Jitsu we had a practice called Reflex training. We train in pairs where one person stands facing the instructor (the attacker) and the other faces away (the defender). The instructor demonstrates an attack, the attacker performs the attack and the defender performs an appropriate defence. The trick here is to practice programming reflexes so that when they occur you don't think about acting you just perform the appropriate defence. This suggests that it is possible to "learn" reflex actions, but these learned actions are a lot more conscious than a conditioned or unconditioned reflex, hence a range of consciousness of behaviours.

This shows that there are a range of locations that we can describe as "reflexes".

In the design I showed Reflexes as a process that links sensory information directly to actions in Figure 27 and Figure 28. The reflexes process is described in section 2.6. Essentially it is a very quick system to activate an action directly from sensory data. So to be clear I am not suggesting that the process described here is located in a single location, as we see in biology there are several potential locations in the brain that reflexes can be generated. While this might sound odd it does not invalidate the model. The important part is that we have several location where reflexes occur in biology, the spinal cord, the medulla oblongata, the amygdala, subcortical structures and potentially other areas of the brain.

Looking at the brain in this way one conclusion is that the brain is almost a layered set of reflex actions of increasing complexity, and at the top is consciousness.

4.3.6   *Memory and Knowledge*

My hypothesis is that the cerebral cortex functions as our long term memory. Working memory seems to be present in the Thalamus and possibly updated by the basal ganglia. While medical technology has established some of the details of neurons and how they operate and the features of the brain where knowledge is stored we don't quite know the details of how this knowledge is transmitted or stored in a functional brain.

Knowledge is represented in the human brain as chemical structures in neurons. As such, the same piece of knowledge will be represented completely differently in two different people. This means that knowledge is difficult to transfer directly. So at present we still need to transfer knowledge by the less efficient means of written or verbal language or direct experience. This also suggests that knowledge is an emergent feature of brains. While this does not guarantee free will, it means that the emergent basis to our understanding of the world hence cannot be predicted purely by physics.

Memories are required for use to operate in a complex world with many objects, names and even complex abstract ideas. The idea is this that to be conscious of something we actually need to have a template, a memory of what the object actually is. This memory is then "loaded" (for lack of a better term) into working memory then matched to incoming stimulus. Most likely we will not even be aware of our memories until they are loaded into working memory. At that point they can be compared to sensory information.

My feeling is that this process explains a few features of human behaviour. I remember after my first cat died *{I miss kitties}* I would often come home and think that I saw my cat in the corner of my eye. I have heard a lot of stories about other people thinking that they have seen pets or even friends or family member who have either died or moved away. My guess is that this is a memory template that has been loaded into memory so that we expect to see whoever or whatever it was. My guess is that this may be the basis for stories of ghosts. People actually perceive someone they expect from a common scene, but when they realise that the person is deceased their perception changes and the person or animal disappears from their perception.

This also happens occasionally with seeing unexpected objects or animals. In the California USA people occasionally see wild animals, such as coyotes or wolves roaming around the city. At first glance these look like dogs. So people have actually reported seeing what they first thought was a dog, then realised it was actually a wolf. The way this works is we actually perceive a common template, a dog, then we think for a moment and realise their perception is not quite right. Then what we perceive actually changes into a better match, i.e. the wolf, and people sensibly freak out a bit.

4.3.7 *Learning*

Children's brains grow to maximally connect the synapses between neurons. During adulthood the connections between the neurons decay where they are not necessary. The same function can be seen in artificial neural networks when decoding the decision structure. What you see is that the network can be simplified to a much simpler structure and still achieve the same decision output. These changes show the learning process. During the maximal linking between neurons the brain is learning the relationships between concepts. During the pruning stage the expertise is maximally understood and the brain is attempting to reduce the energy required to run the network.

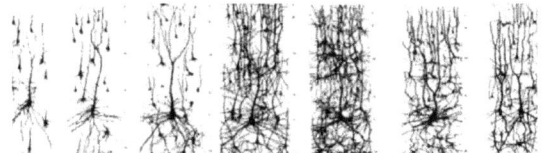

Figure 91 Neuron Synapse connections, Synapse growth and pruning.

As mentioned previously in the section on sleep; the current understanding of how the mind works is that sleep is an important part of learning. There has been a lot of research on sleep memory and learning [121]–[124]. We know that when sleep is disrupted or removed entirely people have difficulty with transferring short term memories to long term i.e. it interferes with learning. Despite the research we still don't have a complete picture of how learning actually occurs. We know that learning requires us to acquire new knowledge and integrating it into existing knowledge structures.

This leads into the difference between knowledge and understanding:
- Knowledge is more or less facts about the world.
- Understanding as we have previously defined is active. It allows us to relate different pieces of information and how changes occur.

When you have an understanding of something simple say a door.

You probably know a few things about doors:

1. They exist.
2. They can be open or closed
3. They often have door handles
4. They can be locked and unlocked

Having an understanding of doors means that you know if you want to get from one room to another you have to open the door. If it is locked you can't open the door.

So learning is both acquiring new knowledge about the world and creating a new understanding of how the new knowledge relates to old knowledge and how changes occur. What we don't know is how this is represented in the brain. This highlights one of the difficulties in biology. We may be able to measure the activity between neurons ad ultimately areas of the brain, but we do not know exactly what information is being transmitted or where calculations are actually occurring.

This is why we need a "functional" model of the human brain and the mind in general. By functional I mean a mapping of the functions that occur in the brain, not necessarily a model that functions.

4.3.8 Theory of mind: Simulating other people

Psychology research has shown that humans develop what we call a theory of mind at between 4 and 5 years old [125], [126]. The most basic experiments typically involve the child watching a sequence of people putting objects in and out of a box, then predicting what a person would see when they open the box. This process allows us to develop diverse desires, diverse beliefs, knowledge access, false beliefs, and hidden emotions.

This means we can look at another person and make a guess at what they can see, and eventually what they might think. It allows us to empathise with others and form social bonds. This allows us to cooperate and form hierarchies, communities and societies.

The ability to communicate including language is something of a requirement for a theory of mind to develop. With language comes the ability to generate abstract concepts which allows us to perform increasingly complex tasks.

In the design terms the theory of mind allows us to simulate other people. For example, if we know a few details about someone, what they like and dislike, it is possible to take a guess to pick a gift that the person will like. We can also guess at what our own social status is and take actions to improve our standing in the society. This is also relevant to what we mean by a simulation. It means we can create predictions about the effects of our own actions on others and others actions on ourselves.

4.3.9 Side track: Structure of digital brains

We can look at the operational structure of a computer and see something interesting parallels to the structure of biological brains. A typical CPU architecture is shown below.

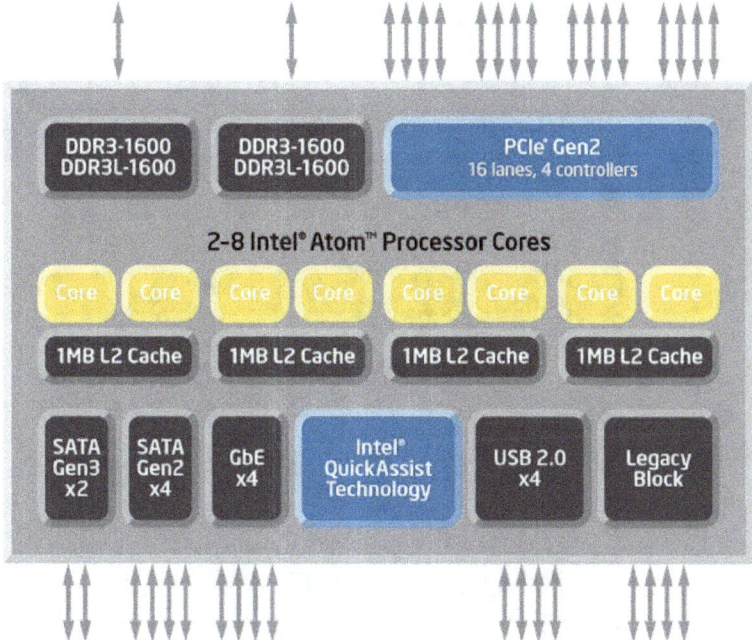

Figure 92 General CPU architecture

CPU memory

Registers might be the closest equivalent to working memory in biology. Registers are the lowest level of memory that the Algorithmic logic unit (ALU) actually performs operations on directly. The next level is "CPU cache", essentially memory that is on the CPU microchip itself. There may be as many as 3 levels of cache depending on the architecture of the CPU. Then there is "external RAM" which many are familiar with from looking at PC specs. That's the 16GB 32GB 64GB up to 128GB RAM. Finally lowest level of storage are hard drives or

SSDs, which would typically be attached to a dedicated i/o bus (typically SATA, occasionally SAS or SCSI). Dedicated Memory chips look like the following image.

Figure 93 An Actual memory microchip

A typical Graphics card architecture is shown below.

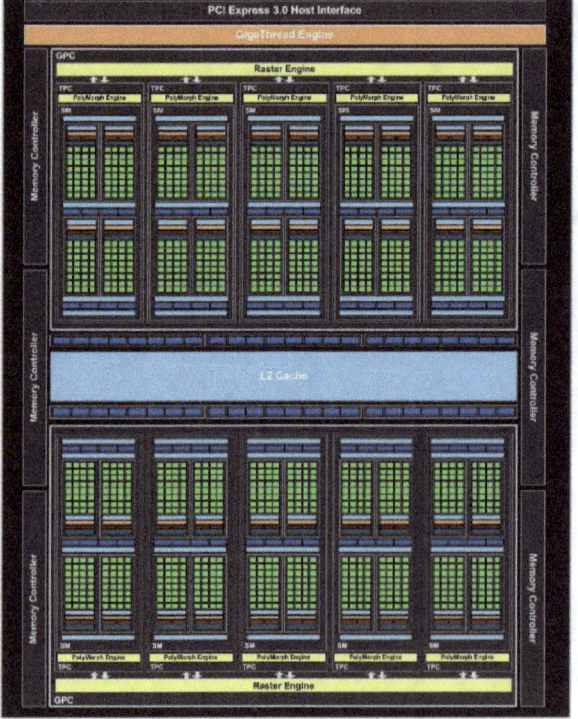

Figure 94 GPU architecture

One thing to notice is that the memory takes up much more space than the actual CPU's which might be consistent with my hypotheses about the cerebral cortex.

The CPUs manage low level tasks somewhat equivalent to the hypothalamus.

Memory in Biological brains is somewhat distributed but access is managed by the temporal lobe.

Graphics cards (GPU's) are a bit more interesting as they have many smaller CPU's, as many as 1000 or more in modern cards, between blocks of graphics memory. The graphics CPU's each run independently in parallel. They are primarily used for taking 3d and "texture" data and projecting them onto a 2d array equivalent to the monitor output. Modern graphics CPU's are capable of performing general purpose calculations and in recent years have been used with "Deep Learning", or neural networks for machine learning applications. This is roughly similar to the Frontal, Parietal and Occipital lobes.

What I find interesting about the general structure of computers is that it is similar in some ways to biological brains. Memory seems to be distributed and close to areas where it is processed, so the equivalent would be graphics cards which has many processors each with its own memory. I suspect a closer analogy would be a GPU with a hard drive along with memory.

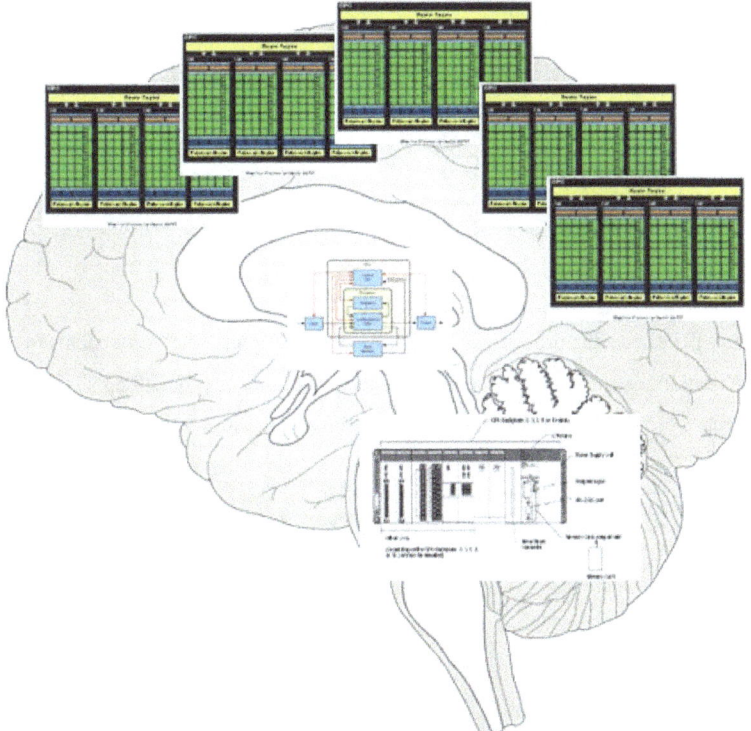

Figure 95 Human Brain overlayed with equivalent artificial components

The midbrain seems to be where most of the interesting things occur, so the guess is that this is the equivalent of the CPU. The equivalent of the Cerebral cortex I think is something like a set of graphics cards possibly with their own hard drives as storage. The Cerebellum and brainstem seem to operate as a PLC or programmable logic controller. This is a low level control device that allows a wide variety of input and output devices to be connected and is capable of performing control system processing.

One interesting idea here is that we use graphics cards to perform neural network calculations. As GPUs have many processors they operate in parallel, and as such it is possibly to use graphics cards in parallel to combine their performance. Each GPU has memory associated with it on the microchip, which is similar to how the cerebral cortex operates.

One major difference in this idea is that I doubt that the main calculations actually occur in the midbrain. The way the CPU actually works by connecting different dedicated functions and allowing them to connect to memory at both input and output via address gates, then the algorithmic logic unit can connect the different circuits and select the required operation. The dedicated factions include addition, subtraction, multiplication, binary masking and many others as shown in Figure 96.

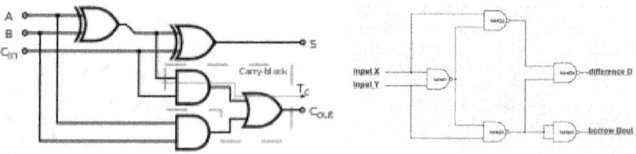

Figure 96 Logic Gate architectures for addition and subtraction

The CPU selects the process with the program counter and selects the memory input and output locations as shown in Figure 97. I suspect the brain does something similar, except that it connects a program running in the midbrain with a program stored in the cerebral cortex.

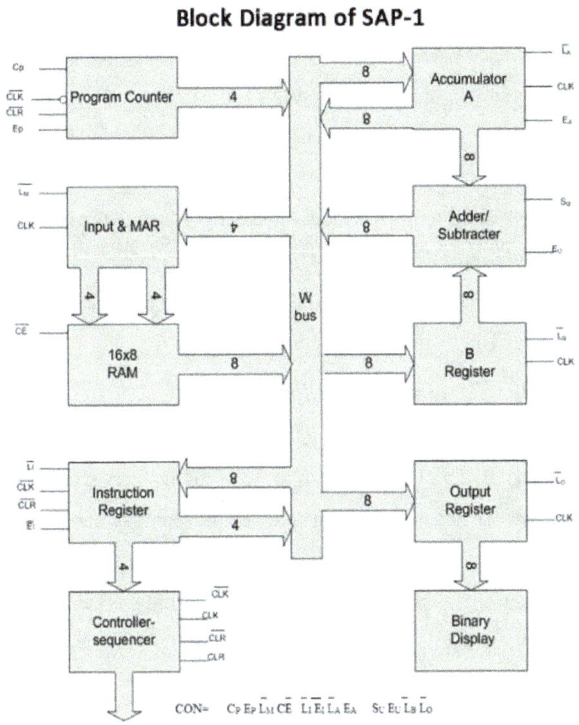

Block Diagram of SAP-1

Figure 97 Block diagram of a CPU

The equivalent mechanism in deep learning is connections between neurons to activate them. What we can and have developed is a neural network that "decides" which process to select.

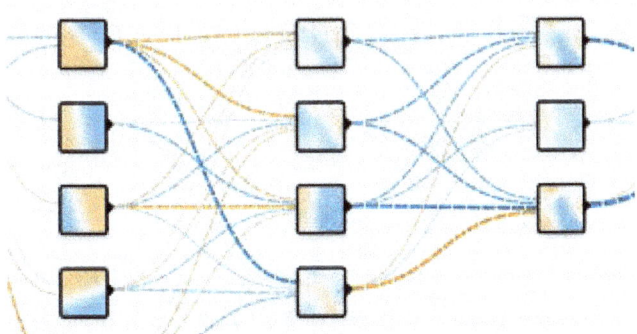

Figure 98 Deep learning layer connections

The analogy doesn't push much further in terms of architecture, but we might be able to get a better understanding of what the biology is doing by comparing how the biology and artificial computing are similar and how they differ.

In terms of power usage silicon based technology is significantly less efficient than biological brains. A human brain might use 20W whereas the equivalent neural network would be running at 20 kW. In terms of calculations CPU's and GPUs can perform calculations many thousands of times faster. GPU's are setup to perform calculations among multiple processors in parallel. As opposed to the CPU which may have six to eight processors on a single chip graphics CPU's may have thousands.

4.3.10   Serial vs Parallel calculations

As it turns out neural networks are quite good at running in parallel. This allows visual data to be analysed in parallel and may explain some of the features seen in research. Specifically the idea that we can subconsciously detect different symbols and examine each to a known template.

Again in computing GPU's have been applied to performing as neural network systems. The advantage they have with this is that with multiple GPU processors these calculations can be performed in parallel.

4.3.11   Energy efficiency

Biology is noted for being low power perhaps 20 W, where as a dedicated neural network processor in operation may use several kilowatts.

Why? Is it just the fact that silicon is less energy efficient or is there a structural different.

So we know that a neural network simulates multiple layers of neurons arguably similar in nature to biological brains. One difference is that in artificial systems all of the neurons are active at the same time. In a biological system the neurons should be able to use energy only when they are active.

4.4   Bugs in Biology

{No not crawling things! The programming bugs!}

The goal here is two-fold, the first part is to look at these conditions and see if any conditions match up to the design and see if we can directly represent any of these problems. A bonus feature is that we might be able to use the design as a model for creating a solution or at least getting a better idea of the problem.

The second reason is to validate the design as a model itself. If these bugs can be represented in the model it means the model is doing something sensible. It might not be complete, but as long as parts of the model match up with the condition then this actually validates the model. Previous researchers have discussed a few disorders that can be explained further using the design of consciousness, specifically Schizophrenia and Alien hand syndrome.

4.4.1   Schizophrenia

Schizophrenia is a problem in society for many people. One of the symptoms of such mental problems is hearing voices or seeing visual hallucination that are interpreted as real. This is often caused by stress and sometimes simply occurs in people.

Dehaene[94] discusses Schizophrenia, showing the long distance connections between neurons become disrupted which disrupts people's ability to compensate for errors in sensory reconstruction. Dehaene "As a result, they would continually concoct far-fetched interpretations of their surroundings".

My guess is that in terms of the information processing these long distance connections my represent the transmission of information between the levels of abstraction as shown in Figure 99. Either the transmission itself or the qualia matching system could be faulty. Either way the objective is for the model to represent Dehaene's description as closely as possible.

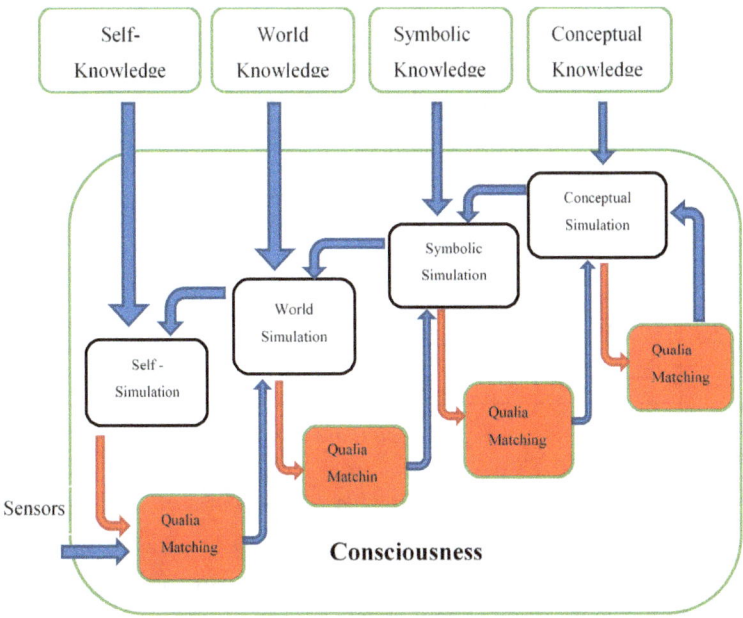

Figure 99 Consciousness Simulation for Schizophrenia

This demonstrates is not necessarily how Schizophrenia occurs in biology, but it shows how a similar problem would occur in an artificial system. If nothing else it shows that this modelling technique can be used to demonstrate existing problems. Perhaps it could be used as the basis for researching treatments or teaching, but that will require research from the psychology and medical communities.

4.4.2 Alien hand syndrome

Alien hand syndrome [70]–[75] is a known mental issue where a person's limb, typically their hand or arm, performs controlled movements, such as stirring a spoon in a cup, without the person directly thinking about it. This gives the person the impression that there arm or hand is being controlled externally i.e. by an alien. The explanation for this is that multiple simulation/control loops could occur in the design as shown in Figure 100.

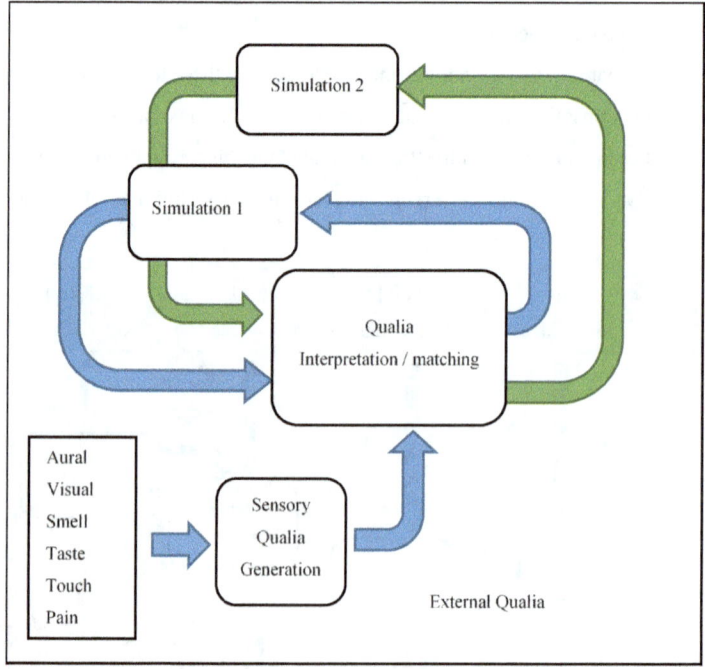

Figure 100 Alien Hand Consciousness loop

This design layout shows that there is two control loops that operate independently. In some ways this is similar to what is described in Section 3.1.12 about "super consciousness". It gives me the impression that a hive mind may not be the most comfortable way for individuals to operate as (if it arose) individuals would often receive stimulus from unknown sources, which could cause frustration or panic.

4.4.3   Blind sight

Blind sight is described as when a person (who often has a brain injury) is unable to see an object but can still correctly describe it when "guessing". My guess is that for some reason the simulation does not reproduce an image. It could be a problem with detecting template objects, or being able to reproduce the templates into working memory where the simulation can use it. Or the simulation itself does not accept or cannot use any detected objects. However, the sensory signal still gets to the brain and analysed, so they might be able to guess what is in front of them even if the simulation says they are not seeing anything.

Almost the opposite of blind sight is from an old joke is "Domestic Blindness". The idea being that people cannot see something that is straight in of them. We have all had this problem, where we cannot see something that is straight in front of us.

The idea in both of these cases is that we cannot completely see something until it is simulated. With blind sight, the object might be running through the simulation, but not being matched and domestic blindness the object is not in the simulation at all so we don't even see it as our brain is not triggered to look for it.

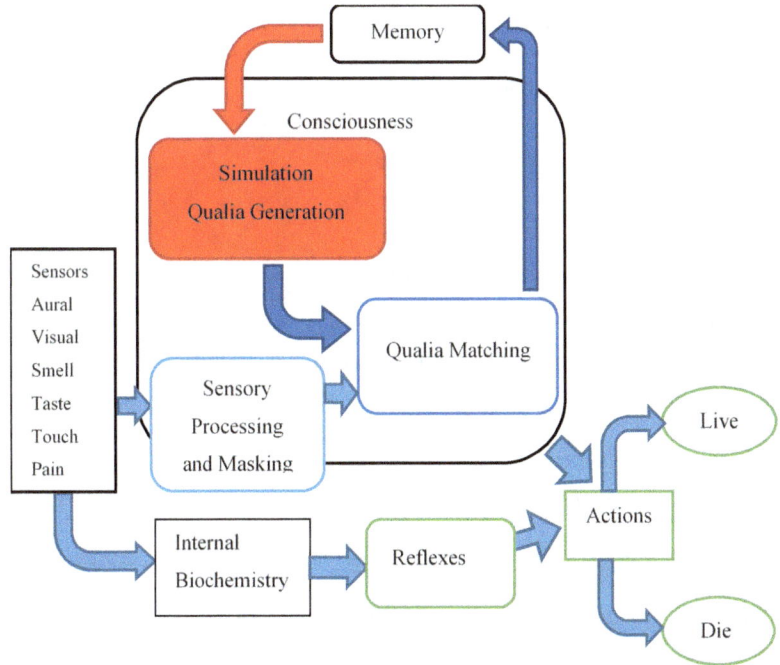

Figure 101 Blind sight

4.4.4 Sleepwalking

Typically during sleep our actions are suppressed, however during sleepwalking people can seemingly "wake up" and perform activities during sleep.

The model shown in We need to suppress or completely mask external sensors during sleep. Figure 102 suggest that sensor processing, consciousness and actions are typically suppressed. However, we know that during sleep we can dream, and often lucid dream, so we know that our minds are still working even when unconscious.

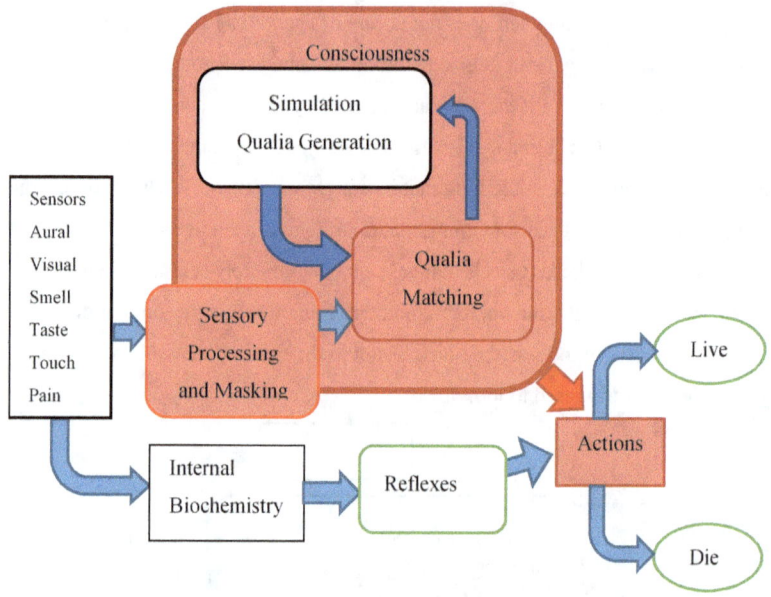

Figure 102 Sleepwalking

4.4.5 Apophenia and Pareidolia

Apophenia and Pareidolia a both phenomenon related to recognising patterns in the environment that are typically unrelated. It may have developed to detect potential predators.

https://www.merriam-webster.com/dictionary/apophenia

"The tendency to perceive a connection or meaningful pattern between unrelated or random things (such as objects or ideas)"

https://www.merriam-webster.com/dictionary/pareidolia

"The tendency to perceive a specific, often meaningful image in a random or ambiguous visual pattern"

A research study using fMRI scans, "Neural mechanisms underlying visual pareidolia processing: An fMRI study" [127] showed:

"Under both the real-face and face-pareidolia conditions, activation was observed in the Prefrontal Cortex (PFCX), occipital cortex V1, occipital cortex V2, and inferior temporal regions. Also under both conditions, the same degree of activation was observed in the right Fusiform Face Area (FFA) and the right PFCX. On the other hand, PFCX activation was not evident under the real-face versus face scrambled or face-pareidolia versus pareidolia scrambled conditions."

During Real face detection high brain activity is found in the Occipital cortex, the Prefrontal cortex and the Fusiform face area. The Occipital cortex is highly related to visual processing, so it is not surprising to see to see high activity here given that this is a visual phenomenon. The Prefrontal cortex is related to reasoning and higher level cognition, which also matches up with the idea of recognising a complex object. The Fusiform face area is a part of the human visual system that is specialized for facial recognition. We know that this area of the brain is active during facial recognition, but we don't know the exact "algorithm" being implemented by the brain. However, we can take a guess based on the algorithms we know in AI, Facial recognition algorithms use a template (a flexible configuration of eyes, nose, mouth etc.), and feature detection to scale orient the template.

This is actually a normal function of the brain. My guess is that we almost always attempt to make connections between stimulus and using a template it can match to "random" features or noise.

The point is I suspect that Apophenia and Pareidolia are due to "hyperactive" template matching of a normal function of the brain. Not dangerous, but an interesting connection.

4.4.6   Optical Illusions

The fact is the brain is easy to trick. My guess is we are easy to trick because most of our perception is based on "object" or "pattern" recognition. All our brain is doing is picking a template we recognise in the image. Hyperactive matching is also described in Chapter 4.4.5 about Apophenia. In many of these images they are deliberately created so that there are multiple templates that can be found in the image or a "bistable percept". Usually people can't see both interpretations at once, but it is possible to flip between them [19], [21], [82], [94], [128], [129].

There are many examples of this from the Necker cube:

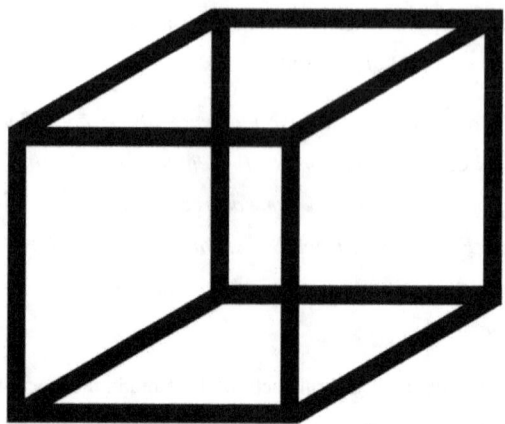

Figure 103 Necker Cube Optical Illusion

The Necker cube can be interpreted as looking down on a cube or looking up at a cube depending on which of the two centre corners are focused on. We have a template of a solid cube which can be matched in two locations. The template matching can only occur in one location at a time, due to limitations in working memory. So even if we know there are two cubes conceptually we match only one cube at a time.

Figure 104 Penrose triangle

The Penrose triangle is a similar concept, where we have a template for right angle corners, but when three are linked together we get confused when we start to consider which corner is in front of others.

The next type of illusion is to consider multiple interpretations from positive and negative space.

Examining Biological Consciousness

Do you see a musician or a girl's face?

Figure 105 Multiple embedded image Optical Illusion

If we look for a shape in the dark areas we can match it to a cartoon man playing a saxophone as a template. Also if we look at the light areas and think of the black areas as shadows we can identify a woman's face. So again we have a conflict between two possible templates and we need to choose one at a time to match to the image.

Then there are other illusions that rely on the lower level visual details such as the following.

Figure 106 Parallel lines Optical Illusion

If you look carefully the lines are all parallel. The impression that they are not comes from the shift in the pattern and the grey lines separating the rows.

Examining Biological Consciousness

Figure 107 Moving Circles Optical Illusion

These use the limitations of our optical resolution and templates again to give the impression that the lines are not parallel or that there is movement in static images.

My guess is the same thing is occurring our brains find a template that matches, but from minor variations in cues our brains give us the wrong match, which it can attempt to correct.

This all goes to show that the brain is easy to fool… but it can be fun.

4.4.7   Binding Synchronising sensory inputs

The "Binding problem" was identified by Eagleman [23]. The problem is that we have sensors at different locations around our body. Touch sensors all over our body, and complex sensors such as eyes, ears and nose almost directly embedding in our brains. This means signals must travel different distances and often have different processing requirements. As such, the senses would not be synchronised and the brain would need to perform a few tricks to delay fast signals. This leads to sensor lag which means that our consciousness actually lags behind sensor inputs.

In Chapter 2.10.1 a system is described as to how this can be addressed in a synthetic system, but we don't know how this is achieved in biology. Research has also shown that there can be significant lag in awareness which may be due to binding related processing.

Eagleman has also shown that we can make decisions as much as a second before we act. One way this might occur is that we actually decide what action we want to perform, then "hold" this action in mind while considering alternatives or simply waiting for a trigger to release a move. Gamers have been measured reacting to onscreen stimulus as fast as 17 ms.

As such, we know that people can react quickly, but this is likely a "held" but pre-planned action, such as tapping a mouse button when we see an appropriate target.

My suspicion is if we need to make a complex action such as determining how to grasp an object to pick it up, a processing delay would still occur as we visualise the grasping pose. In contrast a simple action, such as a mouse tap, could be readied to release at a moment's notice.

4.4.8 Out of Body Experiences

According to accounts from people who've experienced them, out of body experiences (OBE's) of involve a feeling of floating outside your body, an altered perception of the world, such as looking down from a height, or the feeling that you're looking down at yourself from above. OBEs typically happen without warning and usually don't last for very long.

This phenomena represents a variation in self-consciousness, specifically of self-perception and the relationship between the self and the world outside the self.

Olaf Blanke et. al.[130]–[134] has researched the phenomenon of out of body experiences. This research has linked the OBE's with the temporo-parietal junction on the right side of the brain. The Inferior parietal lobule is associated with sensorimotor integration, spatial attention and visuo-motor and auditory processing. And the Lateral temporal lobe is associated with language comprehension, hearing, visual processing, and facial recognition.

These regions are also close to areas that are part of the default mode network.

From "the design" we would expect that the ability to map a person's body and their environment exists in the biological brain. In AI a process called Simultaneous Localisation and Mapping (SLAM) is used to combined the external sensory information into a coherent model. Sometimes this fails spectacularly which results in the vehicle/robot determining an incorrect position.

The hypothesis for OBE's becomes: When a person's ability to map their sensory inputs and their self-model together the brain may become confused and give an incorrect position of the self in the outside world or completely fail to establish a unified world position. This may result in a subjective feeling that the person is experiencing the world from a different perspective. In extreme cases they may experience "no" perspective where all sensory information is overlapping and incoherent.

This may explain why people feel they have out of body experiences under various conditions, including disease, psychedelics, meditations or even direct stimulation. The default mode network is associated with various experiences when people meditate, and Blanke tested patients with brain lesions who experienced OBE's and managed to reproduce such experiences via direct stimulation. Psychedelics can also produce OBE's, but have a far broader effects on the brain.

4.4.9 Psychedelics

There are substances in the world that notably effect people's perception of the world and themselves. So, what is occurring when we take psychedelics?

Areas of the brain that are affected by psilocybin include the visual processing centres including the thalamus and anterior and posterior cingulate cortex [135]–[138]. The thalamus processes and transmits movement and sensory information to and from the cerebral cortex and brainstem, it also regulates sleep, alertness and wakefulness.

The anterior and posterior cingulate cortex is usually considered part of limbic system, which is involved with the formation of emotions and processing, learning, and memory.

Some researchers have shown that many psychedelics affect the default mode network, which may result in feelings similar to meditation or self- contemplation.

Other studies have shown that LSD affects the parahippocampal gyrus and retrosplenial cortex [139]. The parahippocampal gyrus surrounds the hippocampus and is part of the limbic system, it plays an important role in memory encoding and retrieval. The retrosplenial cortex is located and has a function closely related to memory and navigation.

We know that psychedelics can allow people to hallucinate and many users speak of changes in their sense of self, describing a feeling of connection with the rest of the universe, or feelings of understanding or connecting of ideas.

This suggest that there are several areas in processing that are effected. Psychedelics effect the thalamus and cingulate cortex which are associated with sensory information, alertness, wakefulness, emotions, processing, learning, and memory. These effects are consistent with the experience people report when using Psychedelics.

From considering the design, these areas would be sensors and sensory processing, the simulations and our knowledge bases as shown in Figure 108. This includes visual processing, which could be both in terms of generating signals from the eyes, making our recognition system hyperactive or the generative system hyperactive.

All three of these areas (sensors, simulations and knowledge) would affect how we see the world. If our sensory information is being modified, we would most likely experience random patterns that are more or less incoherent. If our recognition system becomes hyperactive we might find that we recognise connections between visual images more common. If our sensory generation system becomes hyperactive we would likely generate our own internal experiences. If any of these experiences are combined we could start experiencing abstract images that could be connected to our existing understanding in the world, then internally generate related imagery. This also seems to be similar to the experience people have.

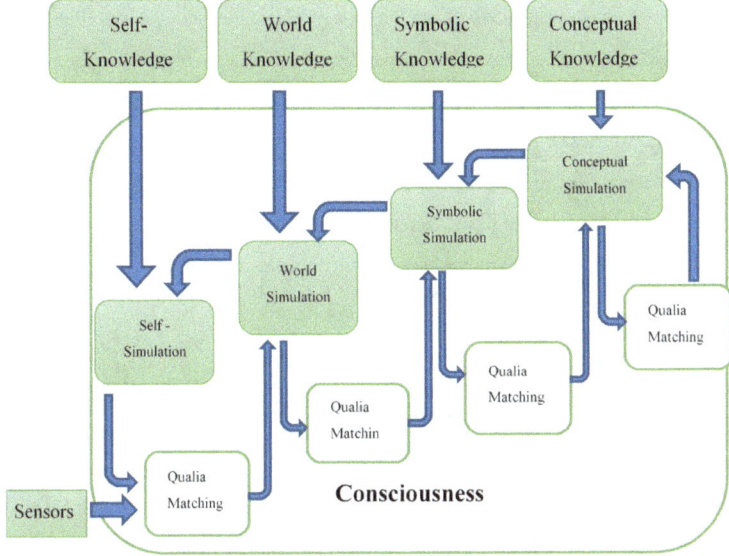

Figure 108 Processes potentially affected by psychedelics

For people to have modified feelings of self our self-model would need to be affected. This could be as simple as a change in the position we understand our senses to be in. If this was changed we might experience feelings of being disembodied. It might cause feeling associated with synaesthesia where our experiences of our senses cross over, we taste colour or see or smell sound.

When considering our higher level conceptual understanding, if these processes became hyperactive we might experience finding new connections between dissimilar concepts which could produce real improvements in understanding or it may simply create the feeling of understanding. For this to create real connections it would need to connect real concepts that were previously unconnected. This would require two concepts to be activated by initially unknown accessors

I should again declare that I have no special knowledge of psychedelics. This discussion is based on the brain research referenced here and my understanding of how the design of consciousness functions. Whatever is going on with psychedelics and the brain is currently being researched by many other people.

4.4.10 Agnosia

Agnosia is an inability to recognise objects. Sometimes this can be purely a visual problem, others it can be related to other senses as well. It has been shown that people cannot recognise objects but they do have the knowledge of what these objects are and can describe them. It can be caused by brain damage in various locations to the cerebral cortex, and can occur with head injuries or stroke.

From the design we can see that the main areas that are relevant are the knowledge bases and or access to the knowledge bases. This could mean that the areas of the brain affected by damage either contain the knowledge or the ability to access the knowledge.

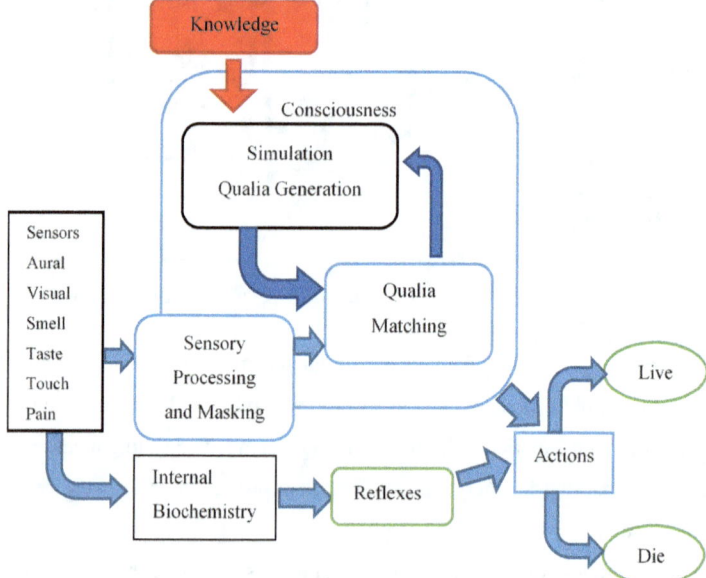

Figure 109 Consciousness with Agnosia

4.4.11 Brain damage

It has been established that damage to specific areas of the brain results in changes in specific functions of the brain. Worst case damage to the brain stem where our autonomic functions seem to be is likely to be fatal because any damage will shut down functions like breathing or heart-beat. We have seen from other researchers that damage to Wernicke's area of the brain resulted in a patient not being able to express words and damage to Broca's area results in patients not being able to understand words correctly.

What we don't know is if this is a processing problem or a "data-wipe problem". Do people lose the ability to process words and connect them together or do we simply lose the knowledge of the words themselves?

Often functions lost from brain damage can be relearned over time. This shows the ability of the brain to change over time, it is referred to as "neural plasticity". This also explain why it is so difficult to pin down the operation of the brain. Most likely two different brain will never operate exactly the same, so we can only establish an average location for any function in the brain.

I am not sure how much this book and the model of brain function could address the problem here. Most likely each individual brain would need to be thoroughly mapped to identify the specific neurons or neuron clusters involved specific actions or knowledge.

4.4.12   Lock in Syndrome

Lock in syndrome has been discussed by a few previous researchers. It can be caused by brain damage and on some occasions by anaesthetics. It is defined by a loss of ability of movement while fully conscious.

In terms of the design we can show this by having some error in output actions, both from consciousness and from internal reflexes. This is most likely caused by interference with the primary motor controls in the brain stem or possibly the cerebral cortex. Either way in the design the links to Actions are definitely effected as well as the Actions and reflexes processes as shown in Figure 110.

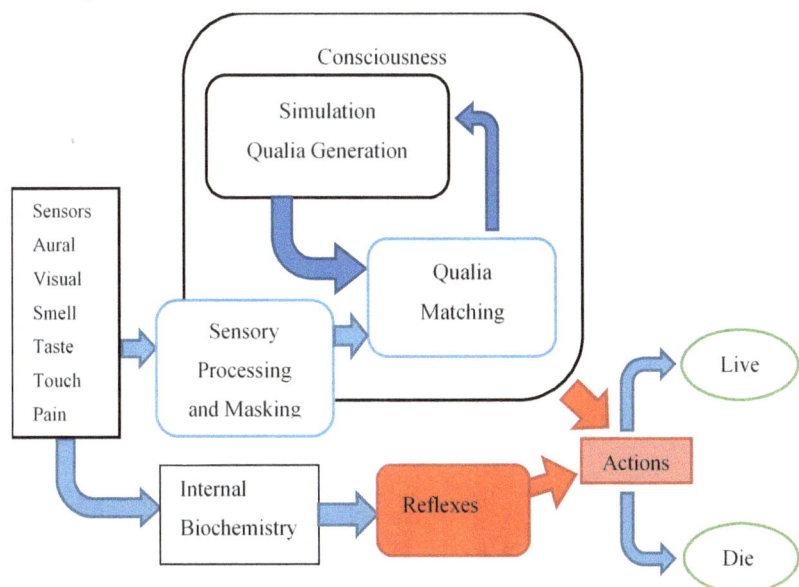

Figure 110 Lock in syndrome

4.4.13 Loss of Consciousness

We know that there are several ways to lose consciousness, sleep being the most common and various vegetative states from disease or injury can be seen as well. From medical research much of this function is relates to the ERTAS (Extended Reticular Thalamic Activating System). This gives us a model of where consciousness processes are occurring in the brain. Unfortunately, as we do not understand what stimulus is being processed and how it is processed it is difficult to say conclusively. The guess I have made is that any part of the consciousness loop could result in unconsciousness, but this is far from conclusive. It will take other researchers work to provide further insight. But if the design is correct it might serve to inform such research.

4.5 In summary

What I have attempted to do in this chapter is examine what we know of the operation of the brain and compare this to how "the design" operates. I see much agreement between the measurements and model of the brain, which suggests that this is on the right track.

The design itself does seem to be useful in describing the process of the baseline operation of the brain and potentially some applications to when the brain fails to function correctly. Whether this can be useful in a medical sense will require someone with more medical knowledge than I have.

Hopefully, I have shown enough in this chapter to explain why I think this is a sensible model.

5 Testing the Design

When practising science it is never enough just to get a solution (even the right one). Any answer to a question needs to be examined.
So for consciousness we can examine what happens in various experiments.

The ultimate question here is: Is this design conscious?

We know from current AI personal assistants, such as Alexa and Siri, which can understand and reproduce human voice. That said, no one has reported their personal assistant attempting to take over the world… *{Yet!}* … So probably not conscious.

The point being computers can imitate biology without being conscious. So we need to look further into this to be able to make a decision.

So… What other experiments can we look at?

5.1 Biological Tests of consciousness

We mostly assume that humans are conscious, but for the most part we don't know from an external perspective we don't know for sure. Several consciousness researchers have discussed Neural Correlates of Consciousness (NCCs). Chalmers who defined an NCC as follows:

An NCC is a minimal neural system N such that there is a mapping from states of N to states of consciousness, where a given state of N is sufficient under conditions C, for the corresponding state of consciousness. [140]

These are signals that can be measured in the brain that show when someone as conscious. Thus far as far as I understand it there are no known signals that guarantee we are consciousness, but there are some signals that indicate states related to unconsciousness, such as REM sleep.

The question becomes does the design give any indication if a NCC exists or where it could be found? Short version no. However this is a current area of research, much of which I am not intimately aware of. As such someone with better knowledge about these signals may be able to identify them more conclusively.

NCC tests determine if the brain is functioning in the way that areas that relate to consciousness are operating correctly. Unfortunately it does not demonstrate what is actually occurring in terms of data processing, data generation etc. So we can't necessarily be sure of why activity in certain areas of the brain results in conscious activity. Using the design it may be possible to find, confirm or possibly refute any such claim, by demonstrating the interactions of such signals with associated processing.

From behaviours we might be able to get some idea at least in people or animals. Something like "indecision" or switching between behaviours might indicate conscious thought in lifeforms, however it could just means a glitch in processing. The good news is that we can analyse what an AI is doing if we can store its state and see exactly what is happening.

Any tests that might show consciousness in biology probably won't work for computers or robotics directly as they are based on completely different signals. As computers don't have neurons there will not be any corresponding correlate.

That said we can simulate the appropriate processing or output behaviours if we know enough about why such behaviours demonstrate consciousness it may still be the equivalent of consciousness in AI.

5.2   Timing of consciousness

It has been shown that it takes as much as 0.5 to 1 second for something to become conscious to a human. I do not find this surprising. The faster the reaction is the more likely the action is directed by a reflex and not a conscious decision.

There are many decisions we make that are based on complex calculations, mathematics in general is something we need to be conscious of to work with it correctly. The outcome of many of these calculations are non-intuitive. In other words we do not get the same result when we think about problems via "human experience" as via calculations.

We think of things being brought to mind, making decisions carefully, with great consideration, "consciously". This means we our thoughts can take a long time to process and often persist for extended periods.

All this goes to the point that consciousness takes time to establish itself. This is not surprising considering the nature of the design. There are several processing operations that occur: feature extraction, object recognition, object template generation, simulation. Even with the efficiency of the brain and parallel processing it would be expected to have some lag between sensory data being received and an object being recognised in consciousness.

As such the timing of any tests of consciousness need to be considered carefully. Our total behaviour seems to be generated by a combination of our "fast" reflexes and "slower" conscious decisions without either of these we would see significant differences in behaviour.

5.3 The Turing Test

The Turing test was proposed by Alan Turing in his paper "Computing Machinery and Intelligence" alias "the Imitation Game ". The general idea of the test is that we can think of computers as intelligent as humans if they can imitate a human without a human being able to tell the difference. As suggested in the introduction to this chapter we have AI personal assistants that can fool people into thinking that they are real people. My guess is, this tests a different thing to consciousness, more or less intelligence.

The other idea is a computer may not have the intellectual capabilities to imitate a human being and yet be conscious. This suggests we need to be looking at a different thing entirely.

Based on all this, my feeling is that the Turing test is not a sufficient test to determine if something is conscious.

5.4 Penrose: Shadows of the Emperor's New Mind

{The emperor got some clothes! Thank oblivion}

A very simplified version of Penrose's argument is that consciousness will not be able to emerge from a calculation. Algorithms are a type of calculation, therefore will never produce consciousness. There are a few counter arguments I can see to his explanations.

The first problem is in Gödel's theorem vs our brain activity. As I understand it Gödel's theorem asks if a program will stop in a finite time with a sensible result. Humans have reflexes of around 200ms with brain wave activity typically between 10 and 40 z depending on the type of activity. This gives us a rough estimate of how fast computation occurs in the human brain. This means that whatever computation or other function that is occurring in our heads it needs to finish in a short period of time, certainly not an infinite amount of time. One thing I feel Penrose has missed is examining the actual "algorithms" that are used in consciousness. Biological creatures do not wait around to do complex calculations. If they waited for complex calculations to complete over a long period of time they would simply be called "Lunch". This means that whatever is occurring in our heads is in no way close to causing a Gödel limited algorithm.

My guess is that we operate much more like a control system. Control systems need to react in "real time", lifeforms in general need to be able to react in seconds or less. Human reflexes can lag as little as 20 ms. Even as a Turing machine, a control system needs to acquire external data which is often chaotic at times. This is a feature I am not sure Penrose considered, at least not in published research. With random inputs we won't know from the Turing machine itself what kind of output will occur.

The quantum effects discussed in microtubules operate at 5×10^{10} Hz, a billion (10^9) times faster than human thought. Penrose suggests this could mean that the microtubules are acting as some kind of computer to get to the same rate as brain waves. This is a lot of computing power operating in every cell of our bodies that has no noticeable effect.

The next problem that microtubules exist in all cells, not just brain cells. So part of the problem then becomes how do we get consciousness from a distributed input and focus these effects in the brain. We know that the brain can shut consciousness down and that if we destroy small parts of the brain we can permanently shut down the brain to produce a vegetative state. So if consciousness was so distributed we would find it difficult to shut down.

Lastly, how do we get specific consciousness features, such as internal vision or speech, out of a distributed consciousness? Consciousness seems to occur inside our own heads, not our big left toes for example. We know that our eyes are associated with image sensing, our ears with hearing, so with a distributed consciousness we might expect a less coherent model of the world emerging inside our own heads.

The other fun part of quantum effects might mean that phenomena such as telepathy, telekinesis or remote vision. *{Yeeeeeeessssss! Cool Jedi Powers!}* Sadly, *{D'oh}* we do not seem to be able to reliably find such phenomena in the real world.

So in both directions, outbound appearance of consciousness and inbound signal processing, Penrose's objective reduction model does not seem to add up ... at least for me.

To consider Penrose's ideas further we can look at how biology works ... or doesn't. Biology often gets things wrong. We jump at "shadows", we incorrectly identify people "I thought you were someone else." Sometimes this is due to bad algorithms and sometimes it is simply due to people not being alert enough. The most basic biology technically needs two algorithms to survive: "Avoid Danger" (Predators) and Eat stuff. For more evolutionarily advanced creatures they also need "Find a Mate" giving three behaviours. There are a few other behaviours that are needed for consciousness, but once we have these we can extend the capabilities with the same structures.

So in the strict mathematical sense Penrose uses biological consciousness is neither complete nor consistent. In simpler terms we can be sure that the algorithms of consciousness do not cover every situation (incomplete) and often make errors (inconsistent). What we can be sure of is that the biological algorithms are effective enough to keep creatures alive in hostile and competitive environments enough for them to survive and reproduce.

The interesting thing about Brains since we are talking about consciousness is that according to Edelman at least, the brain operates as a classifier. In his words: "Memory is an artefact of reclassification of sensory signals". When we consider neural networks and deep learning, they are also indeed often used for "classification" applications. For example what type of object is in this image? What is the best class of action to take next?

Given this we can consider the consciousness program is most likely a "classifier" program that we have already discussed and shown that it is possible to have a classifier examine a program and determine an output in a finite amount of time especially if it allows the set of classes to be finite and or use an "I don't know" class. If we find an "I don't know" class we can either apply a default behaviour (It's moving towards me, I don't know what it is → "Run!") or attempt to determine features of this situation and create a new classifier, thereby giving the system a learning capability.

The idea I have is that rather than running the program itself, we examine the code and determine what type of program is being run, in general terms a "Classifier" program. This is about breaking down the program into patterns. If we find a particular pattern that "guarantees" (in the strong mathematical sense) the program will or will not run infinitely we can create a "class" for the given type of program. For all of the types of programs we find we can add new classes. Which gives us two sets of classes: Finite and Infinite. If we add one other class of program as "I don't know" we can guarantee the program will stop in a finite time. Theoretically there will be less classes of program than there are programs so it is much faster to look at classifications than actual instances. Then all we need to do is run through the list of classes to see which one fits the best and we have a program that runs in a finite time that also has a class. As long as the list of classes is finite this program will stop. If it does not know what type of program it is it will return the class "I don't know".

One of the limitations here is correctness: Will this program be correct all of the time? Entirely possible the answer is no.

So there are two conditions for this program:

1. Will it produce the desired output?
2. Will it be perfectly accurate?

We also know that people who are arguably at least an example of conscious beings, are not particularly noted for being perfectly accurate. In fact people are often noted for being incorrect, or simply downright stupid! This adds up to an argument that we could make an algorithm conscious if we allow some inaccuracy... But I'm not sure how strong this argument is.

So back to the problem where Penrose does not believe that consciousness can be generated from an algorithm. While the mathematical answer might be no, If we allow classifiers to not be perfectly accurate there we can get another answer.

From the design we can examine the requirements an algorithm has to be able to produce consciousness:

1. Data

The program must be able to access data about the external world: sensors.

2. Operations

These are discussed by Penrose in depth with Turing machines, binary numbering systems and other computationally complex descriptions. But they are not enough to satisfy him of consciousness.

From a few others: Koch, Edelman, HPC we also have "re-entrance" or feedback loops.

3. Feedback loops

From my own research what I suggest is that we also need the system to be generative. It needs to be able to generate stimulus internally that can be fed back into its own analysis and correctly identify the type of stimulus being produced. The generative part differentiates this "type" of program from purely analytical ones as we end up with more "data" than we started. If we are generating data, some or most of the data can be discarded without damaging the system itself.

4. Generative system

The system must be able to predict changes in the incoming data. We know we can predict changes in the world around us, from catching a ball, to thinking about what other people are talking about. The general idea is that this allows us to break the deterministic limits of the physical world.

5. Predictive

We may still need to refine what each of these features are to produce a conscious system, but we can say at least that these features must be present in any conscious system.

So the question is: Does this get us beyond the limitations of algorithms to something that can breach the mind body boundary? Or do we still need some other feature to fill any gaps?

5.5 *Revisiting the Chinese box thought experiment*

Testing the Design

The Chinese box thought experiment was originally developed by Jon Searle as an attempt to demonstrate that (as I understand it) basic computer programs are not intelligent. The experiment describes a system where s person is in a room with a "dictionary" of Chinese symbols representing questions and answers. When an input question is given, the person finds the matching question and gives back the associated answer. Does the Person understand Chinese?

Simple answer: No

My question is can the Chinese room be modified in such a way that it does resemble consciousness more closely. Based on the "Design of Consciousness" from this book, the answer should be: Yes.

The main reason that the Chinese box experiment does not produce a consciousness is that it is effectively a simple lookup system with no context as to what any of the symbols actually mean. However, if the person must build the dictionary themselves as a learning process, the answer could be yes! This is because the person must build a model of what the symbols mean as they are added to the "dictionary".

The design of consciousness as described in the introduction must be able to simulate the question that is input into the room. The output then becomes a translation. So with the design of consciousness "the room" or "the person in the room" or "the machine in room" needs to build up an understanding of what the symbols being passed into it mean. To do this it creates a simulation of how the symbols relate to each other.

If it was passed a question in "How are you?" it should be able to examine itself and create a reply based on its internal state.

To accomplish this it needs to do a few things.

1. Translate the Chinese symbols into its own internal knowledge structures.
2. Realise it is a question.
3. Understand that the question is about itself.
4. Examine its state.
5. Formulate a response

Translating Chinese symbols into its own internal knowledge structures takes quite a lot of effort. This is just generally working up a database, with all of the relevant symbols and being able to relate these symbols to other data, images, and models of real world objects. Getting the computer to realise it is being asked a question is reasonably easy: key is the word "How". There are only a few words that are questions and is fairly well established in current AI. Understand that the question is about itself, context of the word "You". Getting a computer to examine its own internal state is not particularly difficult. It can read off its internal sensors, temperature, CPU speed etc. and detect if these values are within acceptable ranges. A simulation can ask the question of itself "Why would I say this?" when it receives any input and work backwards to determine an appropriate answer. All of that mess just to give the reply "I'm fine."

So the question here is again: Is this conscious?

Does adding the knowledge structures make it conscious? Even if we have an extensive knowledge base that allows one set of data (Chinese) to be related to another (real world or English or both). This might make us consider it as "understanding" Chinese, which is (As I understand the problem) the intention of this model. However, I don't think this adds up to consciousness.

Does adding the simulation to the Chinese box make it conscious? If it has a model of itself that it can actively examine inputs and map them onto its own model we could consider it self-aware.

So, my guess here is closer to yes.

5.6 The "Hard" Problem of Consciousness

Chalmers was the person who came up with the idea of the hard problem of consciousness how "inert" matter can become conscious or the mind-body problem.

The idea I have on this topic is that information or knowledge could bridge this gap. Knowledge in terms of humans (and most other life forms) is stored as chemicals in your brain. The information that these chemicals represent is not the same as the chemicals themselves. The information is encoded by neurons and stored as chemistry. But with any encoding system we need to have an associated decoding system to recover the information. As such, knowledge must be an emergent phenomena on matter, chemistry in the brain. From there we can think of consciousness as being emergent from information.

This allows the physical body to be made of "inert" unthinking matter and knowledge to emerge from complex interactions with matter. This can be demonstrated by Conway's Games of Life [6] which demonstrates that shapes can be generated that act like logic gates from a few simple rules. The idea is that consciousness comes from information we actually produce, the internal visions we see or noises we hear in our imagination or dreams. Objects only become a real part of our understanding of the world when we know what to look for, we must be able to recognise them and can generate "images' of them before we can have a conscious experience of the object. This process sits on top of the raw sensory data, so anything we don't recognise either remains as a background "noise" or at times when we are under stress or need to react quickly we might filter out unknown or unexpected objects.

To summarise: If we can accept that information is something that emerges from "states" of inert matter we have the foundation for the solution to the hard problem of consciousness:

Matter allows information to emerge and the generation of information allows consciousness to emerge.

5.7 Features of Consciousness

We have identified several features of "the design" and its operational algorithm.
1. Data
2. Operations
3. Feedback loops
4. Generative system
5. Predictive

Given these features and a general understanding of the capabilities of the design we can go back and examine what previous researchers have considered and see how much actually lines up with their expectations.

Seth et. Al. [4] reviews many features of consciousness that have been examined in previous research. Many of these features are well documented in research. Some are specific to biological consciousness and as such show that an artificial consciousness would not necessarily operate in the same way as a biological consciousness.

Among the conclusions Seth et. Al. [4] shows the following table. This table represents an overview of the knowledge we have about consciousness.

Table 3 Overview of the knowledge about consciousness

Well accounted for	Moderately accounted for	Further development needed
1 Involvement of the thalamocortical core	7 Widespread brain effects	13 Allocentricity
2 Range of conscious contents	8 Limited capacity and seriality	14 Focus-fringe structure
3 Informative conscious contents	9 Self-attribution	15 Stability of conscious contents

4 Rapid adaptivity and fleetingness	10 Subjectivity	16 Consciousness facilitates learning
5 Internal consistency	11 Consciousness is useful for voluntary decision making	
6 Sensory Binding	12 Accurate reportability	

Adequacy of ND in accounting for properties of consciousness. Allocations in this table are necessarily subjective and somewhat arbitrary. However the aim is to provide guidelines for further research rather than a final pronouncement on the theoretical reach of ND.

In terms of the design this can be useful for testing. If we examine how the presented design works in context with these features we can see if the features are present in the design. If all of these features are not present in the design we can say that the design would not be conscious at all. However, if some or all of these features are present this design would be conscious, but not necessarily in the same way a biological entity would be.

These include:

1 Involvement of the thalamocortical core

"As mentioned in the introduction to this article, it is widely believed that consciousness involves widespread, relatively fast, low amplitude interactions in the thalamocortical system of the brain, driven by current tasks and conditions."

The design state:

The thalamocortical core is specific to biological consciousness. As such, it is not relevant to artificial nature of this design. That said this consciousness would operate with a system lag and giving a reliable frequency of operation.

2 Range of conscious contents

"One of the obvious properties of consciousness is its extraordinary range of contents --- perception, imagery, emotional feelings, inner speech, abstract concepts and action-related ideas."

The design state:

The design is capable of generating internal imagery and other sensory data that can then be interpreted by its own internal analysis systems. This could be used to produce a stable state of generating and interpreting sensory data to for what we would consider a range of internal sensory phenomena.

3 Informative conscious contents

"Conscious contents often fade when signals become redundant, as in the cases of stimulus habituation and automaticity of highly practiced skills (Baars 1988). Thus a loss of information may lead to a loss of conscious access. Clear brain correlates have been identified for these effects (Stephan et al. 2002). Studies of attentional selection also show a preference for more informative stimuli."

The design state:

In the design presented the simulation creates an immediacy. The things it would be thinking about would be present in the simulation and as the simulation changes this could represent a loss of information.

4 Rapid adaptivity and fleetingness

"Consciousness is remarkable for its present-centeredness (James 1890; Edelman 1989). Immediate experience of the sensory world may last about the length of sensory memory, a few seconds, and our fleeting cognitive present, though somewhat longer, is surely less than half a minute. In contrast, vast bodies of knowledge are encoded in long-term memory. They are uniformly unconscious. The fleetingness and rapid adaptivity of consciousness require explanation."

The design state:

In the design presented the simulation creates an immediacy. The things it would be thinking about would be present in the simulation. This means again that ideas can be added and removed from the simulation as required to match events that occur in the real world.

5 Internal consistency

"Consciousness is marked by a consistency constraint. For example, while multiple meanings of most words are active for a brief time after presentation, only one becomes conscious at any moment. The literature on dual input tasks shows without exception that of two mutually inconsistent stimuli presented simultaneously, only one becomes conscious (Baars 1988)."

The design state:

When populating the simulation with references to the knowledge base, each reference must be selected. This would mean that knowledge would be stored outside working memory.

6 Sensory binding

"Binding is one of the most studied topics related to consciousness (Singer & Gray 1995; Treisman 1998; Crick & Koch 1990; Crick 1984; Edelman 1993) The visual brain is functionally segregated such that different cortical areas are specialized to respond to different visual features such as shape, color, and object motion. One classic question is how these functionally segregated regions coordinate their activities in order to generate the gestalts of ordinary conscious perception."

The design state:

The design uses multiple layers of analysis to integrate different sensory (Visual, aural smell taste etc) information to be combined into recognisable conceptual and abstract models. These models can then be used to reproduce the relevant sensory data for use in a simulation system where there operation can be considered.

7 Widespread brain effects

"(Baars 2002) cites several bodies of evidence indicating that the neuronal substrates of conscious events have widespread brain effects outside of the focus of current conscious contents, as indicated by phenomena like implicit learning, episodic memory, biofeedback, and the like. Recently the direct brain evidence for this point has grown markedly."

The design state:

The design requires many features outside of the consciousness to be able to function. In fact we can definitively state that the only parts of the system that would be conscious are the ideas or concepts that are directly generated, maintained in working memory and used for predictions.

8 Limited capacity and seriality

"Several aspects of the brain have surprisingly limited capacity, such as the famous "seven plus or minus two" limit of rehearsable working memory (Miller 1956; Cowan 2001), and the limits of selective attention (Pashler 1999). While consciousness is not to be identified with either

working memory or selective attention, it is limited in a similar way, in that there can only be a single consistent conscious stream, or process, at any moment. The serial nature of consciousness can be contrasted with the massive parallelism of the brain as observed directly. Events that occur in parallel with the conscious stream are invariably unconscious. Conscious seriality and unconscious parallelism are fundamental, and constrain any possible theory."

The design state:

The first limitation of the design is a limited amount of memory. To build a machine based off this design would be based on the limited hardware available at present. This may still allow significantly more processing than a human.

The second limitation is that the more concepts are in memory potentially produces a factorial complexity on the interactions between the concepts thus significantly further limiting the capabilities of any such system.

The serial nature of the system is generated from an ability to make predictions of potential future states. These require the ability to consider a sequence of previous events and still maintain space for sensory data being received in real time. .

9 Self attribution

"Conscious experiences are always attributed to an experiencing self, the "observing self" as James called it (James 1890). Self functions appear to be associated with several levels of the brain, prominently orbitofrontal cortex in human beings."

The design state:

A self-model is used to map incoming sensory data and the relationship with external world data. A knowledge of self is required to allow the simulation of the self to accurately predict which actions are possible.

The concepts utilised in working memory are self-generated and as such can be recognised as internal thoughts.

10 Subjectivity

"Philosophers traditionally define consciousness in terms of subjectivity, the existence of a private flow of events available only to the experiencing subject."

The design state:

As Ideas and concepts are generated internally it is possible for the consciousness to produce ideas that are stable but unique to the consciousness.

11 Consciousness is useful for voluntary decision making

Testing the Design

"Consciousness is obviously useful for knowing the world around us, and for our own internal processes. Conscious sensations of pain, pleasure, appetites and emotions refer to endogenous events. Conscious sensory perception and abstract ideas typically involve knowing the outer world. While there are many kinds of unconscious knowledge, conscious knowing may be distinctively useful for executive decision making and planning."

The design state:

One of the main features of this design is that it is possible to gain understanding of events either by experience from a knowledge base or from outcomes of the simulation. As such, this could allow a conscious entity to make decisions based on these outcomes allowing voluntary decisions.

12 Accurate reportability

"It is not at all obvious why conscious contents are reportable by a wide range of voluntary responses. Yet it appears to be fundamental, since our standard operational index of consciousness is based on it."

The design state:

The system would always be able to report what items are running in working memory and any sensory data not currently masked.

13 Allocentricity

"Neural representations of external objects make use of diverse frames of reference. For example, early visual cortex is retinotopically mapped, yet other regions such as the hippocampus show allocentric mapping, in which representations of object position are stable with respect to observer position."

The design state:

Allocentricity is relevant to determining frames of reference of objects. In terms of a simulation it is relatively simple to map out objects in three dimensions with various reference frames or even hierarchical reference frames.

14 Focus-fringe structure

"While consciousness tends to be thought of as a focal, clearly articulated set of contents, an influential body of thought suggests that "fringe conscious" events, like feelings of knowing, the tip-of-the-tongue experience, and the like, are equally important (Mangan 1993; James 1890)."

The design state:

When changes occur in incoming sensory data these changes could be highlighted and used to focus or refocus the attention of the consciousness on the relevant changes.

15 Stability of conscious contents

"Conscious contents are impressively stable, given the variability of input that is dealt with. Perceptually, the confounding influence of eye, head, and body motion is often excluded from conscious experience (Merker in press). Even abstract 'fringe' conscious contents such as beliefs, concepts, and the motivational self are remarkably stable over years."

The design state:
The stability of consciousness is dependent on its ability to analyse accurately the concepts that it produces internally. As such when a concept is produced internally it creates a feedback loop that when accurately analysed will be maintained in the feedback loop, hence producing stable sensations.

16 Consciousness facilitates learning

"The possibility of unconscious learning has been debated for decades, but there appears to be only very limited evidence for long term learning of unconscious input. In contrast, the evidence for learning of conscious episodes is overwhelming. The capacity of conscious visual memory is enormous (Standing 1973). Even implicit learning requires conscious attention to the stimuli from which regularities are (unconsciously) inferred, and the initial learning of novel motor and cognitive skills also requires conscious effort."

The design state:
The ability of the simulation to make predictions could be updated which would allow more accurate simulations to occur. In this sense new skills and information can be acquired or refined, demonstrating the capacity for learning.

From this comparison we can see that many of the features present in biological consciousness are present in the design presented. While each explanation may be tenuous, every feature has some explanation from the operation of this design. This is a reasonable indicator that the design is at least on track with other research.

5.8 *What would they think?*

Dennet: Nope

Caruthers: Maybe

Could supervenience be described as the inverse of emergence?
Emergence is a pattern that is generated from a simple set of rules.
Supervenience means something is dependent on a lower level structure.

Baars: Yes

Edelman: …
Zombie Edelman: Yes

Kock: Yes

Penrose: Probably not

Searl: Arguably No.

Eagleman: Yes

Just guessing…

5.9   In summary

It looks like this design can accomplish a significant amount of what other researchers have discussed as consciousness. There are differences between the design and other theories of consciousness. The design shows a much clearer structure of how information flows through a lifeforms mind and produces consciousness and how the processing determines actions. Other models are typically based off medical information, as such they focus on different aspects of consciousness.

The "tests" of consciousness described in this chapter shows that there are many aspects of consciousness that are consistent with the expectations of other researchers. Some of these tests, specifically the Turing test and The Chinese box tests are more related to intelligence as opposed to consciousness.

So let's look at something I am more familiar with.

{Yaaaaaay Evil robots!!}

6 Artificial Consciousness

AI's can make decisions, learn, generate information, analyse information, but at present we don't consider them to be conscious. We might consider them intelligent.

Considering the biological imperatives, artificial awareness has a major missing component:

Motivation. Biological entities have biological imperatives to stay alive i.e. eat food and continue their species i.e. reproduce (have sex). Robots and computers have no such motivations. Hence they will most likely not develop any behaviours based on such motivations on their own. As such, it is expected that there would be some if not a lot of differences between biological consciousness and artificial consciousness.

There are two main tracks of artificial awareness: robotics and computers in general. The difference between these is the sense of self. A robot has a physical location and form that can affect the real world. A computer is (typically) a box on a desk *{Brain in a jar: A bit like a brain in a jar}*, so the sense of self is far more abstract.

A robot with a physical body could become aware of its position in the world along with what it is. Robotic AI's at present in my opinion are still at the level of a reflex agent. They are most likely not conscious, but they do have some processes in common with Biological consciousness.

A computer has a much more complicated problem as the physical nature is far more simplified. At present on computers a program called a Task Manager runs (at least on Windows™ machines) on computers. The purpose of this program us to show which programs (including itself) are running on the computer and show how much resources (CPU, Memory, Drive access, network access etc.) are being used by each program. From the specification developed this program may be closer to being aware than anything else that exists. Before you start smashing your computer, don't panic as this program has been around almost as long as modern computers and as yet has never attempted to take over the world and destroy humanity. The question we can ask is why? What does the Task manager lack that does not allow it to extend its capabilities and become conscious?

6.1 Robotic Consciousness

While it might be somewhat repetitive, at this point we can make a statement about consciousness in robotics. So…

Could a single separate robot could become consciousness using this design?

If you can accept that this design will actually produce consciousness: Yes.
As we have previously discussed if we have a few features:
- A world simulation
- An internal simulation
- A knowledge base
- Sensor inputs from the real world
- An analysis system that recognises objects from the incoming data, possibly using the simulation that can recognise objects from the incoming data to the knowledge base
- The ability to populate the simulation with objects from the recognition system

If these features can be implemented in a robotic system we can argue that the robot is self-aware and potentially conscious.

6.2 Computer Consciousness

If we can implement our previous features for a robot we can definitely implement them for a computer. However, this idea is slightly different from a robotic simulation as even if a computer has a world simulation it is basically a box on a desk somewhere. The computer mainly interacts with the world via information.

As for internal awareness we can consider a "Task manager" on your computer. It would definitely be able to create a world model/simulation and a symbolic simulation, but the lack of a physical body would change the experience significantly. This would more or less be a brain in a jar.

It would be possible to connect cameras, and other electronic sensors to give it a more general sense of the world. Connecting this kind of AI to the internet could be interesting. It would need to be able to build up a conceptual model of the world and recognise data from itself and the internet as concepts from its knowledge base.

The whole idea of the design is that we have an internal simulation of the world that generates consciousness, similar to how images can be generated. A computer based Brain in a jar AI would be simulating what it expects to see from external information sources. For example this AI might be able to predict what it finds in a web page from a basic description.

6.3 Distributed Consciousness - Hive Minds

Distributed awareness can be described as a hive mind. Essentially, the idea is that a group of non-conscious entities, from bees or bacteria in biology to a collection of web pages in the internet for technology that act as the equivalent to neurons in a human brain. The idea is that somehow these collections can exhibit a consciousness.

So, how would a hive mind operate?

Based on the design a hive mind must have the same functional features to have a consciousness.

From the design these features are:

- Sensory inputs
- A Self-model
- A world model

These features must be able to combine these features into a process that transmits physical feelings, thoughts and ideas and allow them to be processed by the self and world models. These processes can then be fed back into the sensory interpretation system to estimate the state of the entity itself and the world around it.

How can a group of non-conscious entities transmit information in such a way that they can create the models? A group of bacteria can transmit information by chemistry. A group of bees can transmit information by using chemistry and more complex systems such as vision or sound. A collection of webpages, at least at present do not communicate with each other, they communicate via the people looking at the internet. So I think based on this we can rule out that the internet is or ever will become a conscious entity as long as it remains with its current functionality.

With a hive mind there is some difficulty distinguishing between self and external.

6.3.1 The speed of thought

A hive mind of bacteria could communicate but far slower than a human brain. This would result in a consciousness that operates at significantly slower rate. Human reaction time is typically around 0.2 seconds with neuron transmission speed of as high as 120 m/s. The chemical communication of bacteria or bees of a hive mind would communicate at a rate much slower. So if the hive mind transmission speed was 1 m/s the reaction time would be around 24 seconds. This obviously changes how quickly the hive mind would be able to react or even communicate.

Extending this idea we can consider stars galaxies or even the entire universe as a single conscious entity. Obviously the transmission speed for thoughts would be measured in factors of light speed, with millions of lightyears between galaxies, so the reaction time would be orders of magnitude larger.

Taking this idea back to how we would communicate with this consciousness we can come to the conclusion that hive minds may exist at timescales outside our own perceptions. As such we may not be able to see if these consciousness exist and conversely some of these may operate much faster than our own perceptions can operate. As such we can conclude that the consciousness must operate ate similar timescales for "us" to be able to communicate with.

6.4 Artificial vs Biological Consciousness

The biggest difference between a robotic or computer consciousness and biological consciousness is that a biological consciousness has motivations. These motivations come in the form of hormones such as Adrenaline, Dopamine, Oxytocin, Serotonin, Melatonin and many others.

It is possible to simulate the effects of such hormones in an artificial consciousness by using variables to simulate hormone levels and adjusting outputs based on hormone levels. The difficulty with this strategy is that even if this was implemented the difference with the real world is that hormones are chemicals which real with each other and other chemicals in the real world. Many of these interactions we do not currently have accurate information on what these interactions would effect. As such any simulated consciousness would not necessarily react in the same way a biological consciousness would, which makes the whole exercise somewhat pointless.

6.5 What I am not sure about.

The design of consciousness is based on the operation of biological consciousness. It has been developed to mimic what I think is happening inside our brain to develop consciousness, the light going on inside our brains. The main cycle of the consciousness loop seems to be: use the simulation to generate qualia, match simulation qualia to sensory qualia, analyse qualia to produce concepts to populate the simulation. This looks like it holds up for biology.

I am not so certain for artificial consciousness. Computers can create simulations (with some difficulty), generate images or sounds relatively easily, and analyse images to recognise objects in a scene.

Artificial consciousness may be conscious but not intelligent. This I feel is also consistent with biology. Living creatures could also be conscious but not intelligent. Decisions can be made by a variety of mechanisms in computing. This I think satisfies many of the tests of intelligence, but does not necessarily make an AI conscious.

Artificial emotions could be implemented simply by using a variable to describe various hormone levels and allow the hormone variables to be modified by external stimulus and the decisions to perform actions to be influenced by the hormone variables. Easy to implement, and it might simulate something similar to emotions. My feeling is that creating emotions in artificial intelligence is a really bad idea.

{No-one wants their toaster to attempt to usurp their authority.
Toaster: Down with Humanity!}

So all of the aspects that are used in the processing can be performed by artificial means, however I am not sure if this is enough to justify artificial consciousness. There is a shortcut: as the simulation is already active and the concepts within are know it is possible to bypass the analysis stage. This can be just a simplification as the matching process could take place and only analyse if an unknown is detected.

In biology the simulation is aimed at generating "raw" sensory data internally which is then analysed by the same circuits that analyse external sensory data. If the artificial system performed this it might be enough as the emergence patterns are repeated, however this is very inefficient for an artificial system. The question is: Is this system, even optimised, enough to justify "consciousness".

Then a further question is: what is missing if anything?

7 Conclusions – Answering some questions

The most important Questions

From the design we can take a guess at how consciousness operates. When we look at the operation we can answer a few questions. The following section takes a look at a few questions about both artificial and occasionally human consciousness from which I think we can make solid conclusions from.

Much of these questions are about when AI goes bad. More importantly I am interested in looking at exactly which conditions make AI dangerous and what we can do to make sure that the AIs don't kill us all. The good news is that as far as I can tell we are nowhere near the level of complexity and integration that would make AI's particularly dangerous. Whether or not we will get to that point is another question entirely.

7.1 Will computers try to take over the world?

{Evil genius: Muhuhuuhahahahahaha}

The big difference between computers and life forms is the motivation.

Life forms are motivated to stay alive: pain

Life forms are motivated to reproduce: Sex

On the other hand computers don't care!

Order a computer to kill itself and it will with no hesitation. *{Don't try that with a life form.}*

So would an artificial awareness suddenly decide to kill humans? Probably not. However, let's put an evil hat on and take a look in more detail.

7.1.1 Plan A: Actively trying to kill us!

{Evil Genius: Excellent! My evil Robots will rule the world!}

Conclusions – Answering some questions

The Terminator movies have shown a war between humans and machines. Where an AI became self-aware and deciding to kill the humans. For this to occur a lot of things need to happen: Mining, mineral refinement, manufacturing, establish supply lines, actively look for humans and then shoot them, well, us. This means the robots would actually need to control all of these supply lines and coordinate between resources manufacturing and deployment. This is difficult to do especially without being noticed.

{Evil Genius: Damn!}

So not likely to happen unless we build the whole system ourselves. While we are currently attempting to automate mines and other parts of the manufacturing process, the connectivity between processes in still very manual. We sell resources between mining and manufacturing so this is an inherent hold point at any stage of the supply and manufacturing system. So this is not likely to happen … today. If we ever manage to automate distribution and manufacturing, we might have more of a problem of the entire system being subverted.

The whole point is: it will be kind of obvious if the robots attempt to take over with conventional weapons.

In The Terminator movies the robots fired off the nukes first, which could still be a problem. On the up side nuclear launches are probably and hopefully one of the most protected systems we have so getting unauthorised launch access is difficult, but not impossible. The obvious solution there is make sure that computers cannot directly launch nuclear missiles, or better yet decommission the nukes. This still leaves open the idea that some existing piece of infrastructure could be used to destroy all humans.

There are two major sets of infrastructure that we rely on for our civilisation Power and water. These are again still fairly manual, widely distributed and fairly unique, so coordinating a shutdown of these services is not a simple task.

So the key for a direct strategy could be that automation would need to be universal in society. It would require automated mining, manufacturing, supply/transportation, strategic control and communication systems. It is certainly possible to automate these processes, but this still leaves coordinating between these processes. These are usually managed by salespeople. We want the money to transfer along with the goods. So if an AI was attempting to take over we would likely see some effects which would include bypassing the trade system. This would be noticeable as spontaneous transfers of goods without "people's" knowledge would occur. This could occur by an AI "hacking" a sales system or collecting money and making automated purchases.

Conclusions – Answering some questions

The other potential direct attack would be to have universal "helper" robots which could be corrupted, as depicted in I Robot (2004). We would require these robots to be everywhere and have the capability to overpower humans. There are a few humanoid robots around. Honda created Asimo in 2000. A humanoid robot capable of walking around and performing various activities including interacting with humans. In the last 20 years this field of technology has improved significantly and there, however they are still far from universal. The biggest problem with humanoid robotics and indeed any autonomous robotics or vehicles is power or fuel. At present the energy we can supply such technologies with a limited energy supply. Eventually they run out of fuel. The humanoid robots typically only run for a few hours, perhaps a day. At present electric cars have a range of up to 400 km, fuelled cars perhaps twice that. The point is eventually these vehicles run out of power and stop. This limits their mobility and hence the capability to achieve any goals.

The conclusion is: if something were attempting to "take over" or "destroy the humans" we would notice fairly quickly. It might kill a lot of people, but ultimately we would have a reasonable chance of fighting back and stopping any such strategies.

The main point is that with automation we will be setting ourselves up for this kind of failure. "The creator is destroyed by his own creation". Most likely this kind of scenario would be caused by humans automating everything. The technology to do so would have to be everywhere and given enough capacity to overpower our own civilisation.

So the solution to this is to be aware of how everything interacts and make the effort to notice if this screws up somehow.

There are probably a few other ideas I have missed. So most likely if something does go wrong I'll be one of the first to get whacked. *{If I'm not the cause of it in the first place…}*

7.1.2   Plan B: Steam Rolling. Stepping on ants!

{Evil Genius: OK Plan B! Make it look like an accident.}

However the big problem would be that a computer system could get large enough that it could perform some task without considering or realising it was damaging people. These AI's in performing the correct intended actions will decide to do something that is ultimately harmful to the people who use it. For example getting autonomous robots to build highways. If it screws up for some reason it may decide to keep building a highway through people's houses.

Conclusions – Answering some questions

That said we can make a detailed examination of what would be required to create such a system and examine how this system would cause a steam roller event. The steam roller situation would require the computer system needs to be extensive enough to be effective at its task. Again this is not an easy system to construct. As yet humans have not constructed a computer system that can logistically coordinate robots, supply resources actions over an entire country. From the highway example the computer system would need several capabilities:

- To be able to act across the length of the highway
- The planning capabilities to create and modify designs
- The resources (Asphalt, aggregate, sand etc.) and the ability to access the
- "Agents" capable of performing the construction and destruction, Bulldozers, Back hoes, Dump trucks, Steam rollers, Asphalt pavers etc.
- The permission of authorities to begin such a project.

In this example we can recognise that many of these capabilities do not yet exist. It is probably possible to create such a system with the technology we have available at present. We have all of these capabilities separately, but these capabilities have not been combined into a single system. We have plans for building highways, we can automate robots, and theoretically we could automated mines, and logistics systems.

The reason this example in particular would not cause huge problems is that as soon as the robot highway maker attempted to knock down someone's house who hadn't been notified or approved the demolition of their house they would complain. In other words they would notice and as such we would notice. For the steam roller to continue means we would more or less be agreeing to keep it going. So in this case it would be a failure of humans to stop the system from operating one they noticed it was causing a problem.

At a more personal level medical robots have begun to appear to both perform surgery and act as expert systems to diagnose medical conditions. The downside is that Medical AI's may not diagnose a person's medical condition correctly and accidently cause that person, in a worst case, to die. Unfortunately, for us human doctors have the same problem. The good news is that we have medical experts who examine when people do die accidentally to determine if the death was preventable. This would be the same for medical AI's. They could potentially decide to murder someone on the operating table. It could be possible for the medical AI to deliberately give incorrect information and prescribe lethal drugs to a person. The victim would still need to take the drugs. If we have a surgical robot/AI that actually cuts people open to perform surgery, again with the right programing, it could decide to cut a

Conclusions – Answering some questions

major artery or the patient's brain stem. In this case in an individual death a medical examination of the events should be able to determine if the death was accidental or not. The problem is you still need to submit yourself to said robot. As soon as they are established as killing people, most people would stop going to them.

In summary, at present we have not built as system capable of autonomously running a steam roller event, secondly we will notice that a steam roller event was happening.

7.1.3   Super intelligence

Sam Harris gave an interesting TED talk about AI and if we could develop it without killing ourselves. "Can we build AI without losing control over it?"

https://www.youtube.com/watch?v=8nt3edWLgIg

He presented 3 Assumptions

1. Intelligence is a matter of information processing
2. We will continue to improve our intelligent machines
3. We are not near the summit of our intelligence

The general idea of this talk is that eventually we will create a super intelligence that will be able to incorporate us (humans) or steam roll us.

So can a super intelligence be created? Nope. Intelligence is as much a matter of the problem itself as the intelligence used to solve a problem. A 'better' algorithm comes to a solution quicker than a less intelligent or even less informed algorithm. It will not automatically solve a problem outside the problem space. The same can be said for human intelligence. You can't develop a strategy to win at noughts and crosses. It does not exist. It does not matter how smart a person or algorithm is. We know this because small problems such as noughts and crosses has a very small problem space that can be completely examined. It does not matter how smart you are, there is no strategy that guarantees winning.

Conversely we can think about where the destruction or enslavement of humanity is part of the problem space. These would be problems dealing with humanity and the management of the resources we have access to. Problem spaces for the management of countries, politics, global militaries and a few others might have these properties. That said

Even if an AI does go nuts and start killing people, my first instinct would be to check for a rogue programmer. Frankly, this is the same with most terminator scenarios. It is less likely to be the AI itself deciding to end humanity than another human either accidentally or intentionally creating an algorithm that causes a problem.

7.1.4   *Will computers ever spontaneously generate emotions?*

No.

Emotions in human and other animals come from hormones and very simply Computers don't have hormones or any other biological motivation. As much as it might seem like it sometimes, your toaster or phone is not going to try to kill you because it "hates" you.

It is possible to program in emotional features intentionally, but this comes under the previously mentioned unintentional consequences of programming. We are more likely to intentionally implement such a feature to see what happens, then notice that the behaviour becomes inconsistent. The solution here is to avoid implementing such behaviour and make sure that any such computer systems do not have the capacity to kill people indiscriminately.

7.1.5 AI in a box

One of the main concerns Eliezer Yudowski has brought up is about "letting AI out of the box" [141]. The concern is that one an AI is freely able to move between computers and actively participate in the real world we are in trouble.

A defence is: AI is inherently stuck in a box, we call them computers. Even the Internet if you are that concerned is still made of computers so you don't get around it in that way. Once you try to get AI out of the box you get nano-bot grey goo which is another problem entirely. This has one of the big limitations all robotics has: where does the power come from?

Humans are inherently stuck in a box: we call it your head or maybe just your brain.

Attempting to remove a human from their head tends to be messy and not end well especially for the human. But it is an effective way of stopping a human being from continuing to operate, so if an evil human attempts to take over the world, it is always an effective strategy to shoot them in the head. Likewise, AI's inherently run on computers, if they act up we will always be able to shut down the computers.

My point is we are attempting to get AI's to be human, this is a well-travelled and if not well understood problem space, because as we are human we are constantly travelling in it.

We do know that humans have issues as we regularly kill each other, but we also know why many of these wars occur and have put in place measures that limit the possibility of doing damage to each other i.e. democracy, governments, laws etc. I expect AI's will have similar restrictions. That said, if you do see an AI run for president, you can start worrying.

The other way AI in a box can be a problem is with the internet. An AI could potentially modify what we see via the internet for its own benefit. Again the defence is simply to be sceptical about what we see on the internet, use various sources and actually look at the real world.

My feeling with reading some of Yudowski's work [141]–[145] is that he extends the ideas of super-intelligent AI well beyond our current capabilities and potentially considering

Conclusions – Answering some questions

the "end state" of all possible AI development and adding in technologies such as "nano-bots" which are still science fiction at this point. What I would like to see from Yudowski is a sequence of technological developments and events that would lead to a "catastrophic" AI event and the nature of the AI.

7.1.6 Creativity

Will a "super intelligence" be able to out create humans? Yes.

The way AI has currently started to beat humans is by being able to search through the problem space network more efficiently than a human can, not by being more imaginative and creating new regions to examine. One of the best strategies we have would be GAN's (Generative Adversarial Networks) could be modified to create a creative AI. On this theme Google and Facebook developed Chat-bots that began to develop their own language were then shut down the program as it was not achieving the goals they were aiming for and frankly it was a bug not a feature[146], [147]. This is an example of unexpected creative development related to language. A similar structure could be used to hide computing intentions, but again we would notice the AI doing something strange.

In analytical fields we could use GANs to generate and test designs and potentially find faults or improvements that humans miss. This is an idea that shows the beneficial side of this approach. An AI could be developed to explore a problem space creatively and find solutions to a problem that humans wouldn't or couldn't think of. Theoretically, we can imagine a scenario where this kind of AI out creates humans and uses this ability

If this idea were to go bad it might look like a steam roller situation where the AI generates solutions to problems in our civilisation. The limitation here is that the AI would be able to generate potential solutions, but the real world would still be required to implement such solutions. Creativity for things that work in the real world requires experimentation. It is possible to come up with many to an extent infinitely many ideas about how the world could work. Physicists do this every day, it is their job. The problem is just because you come up with an idea does not mean it will work in the real world. So to be creative we need to do experiments in the real world. Which experiments to do might be improved by better intelligence, but hopefully we would notice a robot army building particle accelerators. So when weird buildings are suddenly constructed without humans knowing initially, we can worry.

Very simply: we are not there yet.

Can we get to the point where computers can take over? Yes, but we have a lot of work to create such a system.

7.1.7 Why I'm not worried

{Evil Genius: Damn!}

So what does this all mean? My conclusion is that AI as we know it will not cause the destruction of the human race…

Eliezer Yudkowsky may disagree. I have looked at his work and I am not overly concerned with AI causing problems. However, I do think his concerns are worth examining, certainly at a philosophical level and probably at a practical level. As the saying goes "The stone that you trip on is typically one you didn't see" … so I am always happy to look out for ideas that we might trip on. That said it is not enough just to look for them. What we need are ideas and more specifically tests for how to see if we are going to have a problem with new ideas or technologies.

With the two previous types of Killer AI: Active Killers and Steam Rollers you can think about them in terms of how AI actually works.

- For an AI to actively decide to kill people the decision to kill people must be part of the problem space.
- For a Steam roller to occur the program must be doing something that is a by-product of the output decision but not an intentional decision by the AI.

The trouble comes when the "kill the humans" state pops up in the problem space and we don't notice. This might happen with learning systems, but again examining the problem space itself will show that "kill the humans" state exists. This means that we are not likely to find our toasters attempting to kill us is all we are doing is getting it to make toast. Even our current AIs like Watson (one of the most general AI's we have) is not going to actively decide to kill humans as the scope if its problem space does not include killing people, simply answering questions.

You might get there if we actively attempt to create an AI the actively searches out and kills human being i.e. the Terminator scenario. Again this is a deliberate act by humans to create a robotic system including a manufacturing and supply chain to create weapons and actively targeting humans. Some idiot might try it, but we will probably notice.

The solution is clear though:

- Keep examining the performance and consequences of any AI system created.

- Make sure that <u>anyone</u> can shut them down. If we see examples of an AI doing something unexpected or destructive make sure it gets shut down.
- Make the effort to understand why the system went wrong. When we shut an AI down and investigate why it made the problematic decisions.

This is even more important when we find that the AI system has the capability of accessing tools without human intervention. The more capabilities it has the more chance there is of something going wrong.

7.2 Can consciousness be transferred?

There have been a few people and organisations that are currently attempting to research how to transfer consciousness to either another body or a computer and thus be able to live forever. This chapter will discuss a few of the aspects of this idea and ultimately attempt to answer this question.

From our design the limitation of consciousness is a combination of the internal simulation and the sensory systems attached. So if the Consciousness loop could be expanded to include "new" functionality i.e. another body it would be possible to be conscious off the new body. This requires the simulation to be expanded to include the new body or new components. This expansion is a lot more than just being able to imagine the new components, our mind needs to be able to incorporate the new component and combine it with its existing structures. This means it also needs to be able to receive information from the new component and be able to relate the inputs with the component itself and existing structures.

The problem is if you separate any physical forms and the internal communication and the sensor inputs, the consciousness would be separated as well. When we do separate a component from the consciousness we get something like Phantom limb sensations with amputees. Essentially the simulation still thinks that the component is attached and ends up continuing to "feel" the limb after it has been removed.

So we can expand this idea to complete people, sharing consciousness, human robotic/computer consciousness, and then finally transferring consciousness. Based on the design sharing consciousness requires two consciousness being aware of each other. This means each consciousness must have a simulation of both them self and the partner. It also means they must have shared inputs. Anything that is felt by one is automatically felt by the other. This would mean that both consciousness have the same model and the same inputs. Eventually, what <u>could</u> happen is that the simulations become so combined that they become a single unified structure.

Conclusions – Answering some questions

If both consciousness are synchronised they would in theory be fine. However, if the consciousness are not synchronised it could create a problem that an action performed by one consciousness may not be expected by the other and we end up with a phantom limb situation again.

There are some interesting phenomena in people called alien limb syndrome where a limb, typically a hand may move on its own accord for the person it is a part of. I suspect what causes this is a form of "separated" consciousness. This is almost the opposite of a combined consciousness. Essentially typically through some form of head injury part of the brain becomes isolated. This part of the brain must have access to the limbs "outputs", its ability to make the limb move and an area of the brain that can simulate the limb. As such what occurs is the limb will move on its own in ways that the main consciousness does not expect. Hence the limb operates without the input of the main consciousness.

7.2.1   *Human-Robotic consciousness*

Humans and robots can have much of the same physiology. By this I mean arms, legs a head and as such humans are able to understand the difference in simulations. This is shown in our ability to use prosthetic limbs. The hypothesis we can develop from the design is that we can become "conscious" of any system we receive sensory information from. Once the sensory information is received continually our self-model can be modified to take the new limb into account. Similar effects can be found in various areas of expertise. When learning Swords one idea is that the sword becomes "an extension of your own arm". Pilots operate an aircraft or even experts playing a musical instrument. In effect the "tool" does become an extension of our own self model thus allowing us to create expectations of performance far faster than if we simply consider ourselves operating a tool.

In these examples humans expand their self-model and hence self-consciousness into the realm of robotics.

7.2.2   *Human computer consciousness*

Computer consciousness is another level of abstraction from humans as they do not have the same physical components. However, with robotics we are still creating a computer simulation, so it should be possible to create something equivalent to a computer. The question is what is actually being simulated in a box on a desk. The answer I suspect is the internal thought of the computer. As examined in chapter 6.2 something quite like "Task Manager" on windows that shows which processes are occurring. Humans may be able to learn to modify their self-model to take into account alternative processing methods in the same way they can learn a new tool with their hands. If they have a brain computer interface a person may be able to link to an abstract concept that allows direct internet access. This also means the self-model can be adapted to take into account mental capabilities the same way we might learn how to perform a new mathematical function.

7.2.3   Telepathy – Shared Consciousness

Now this might be a weird thing to put in a book about consciousness... sort of. We can think of telepathy as experiencing someone else's consciousness. For telepathy to work a human brain would basically have to operate as a fractal antenna that reads a holographic image of another brain. The idea being that a brain would produce a signal that can be projected or reprojected in three dimensions (a hologram). This signal would need a complex antenna, a "fractal" structure that can detect and reconstruct the 3d signals from one brain and map these signal to another brain.

Since each brain is a unique combination of connections between neurons and data presumably stored in neurons, the mapping needs to be fine enough to map between the differences between each brain.

For example we can all understand the word "Word", however where "Word" is stored in the brain (presumably in or near Broca's area), if it is not distributed storage which would make the problem much more complicated. Creating a mapping between brains is an incredibly complex task. That is not to say it is impossible. Researchers have managed to wire two people together to allow them to interact via a direct brain interface [148], [149]. While cool, this research shows the difficulty involving the decoding signals from one brain and allow these to be encoded for a second person. But from what I understand we need to improve the bandwidth of what can see.

"This paper presents the successful demonstration of a new, non-invasive BBI in humans, which allowed pairs of participants to successfully collaborate and complete a series of question- and-answer games using information transferred between their brains. This BBI paradigm significantly extends and improves previous protocols in that (1) it involves the transfer of consciously perceived information in the form of phosphenes, (2) works in real-time, and (3) permits bidirectional information exchange between two participants."[148]

What would this feel like?

I can come up with about 3 ideas of what this would be like.

1. Alien Mind - we feel a bit like how people describe Alien hand syndrome or People with split brain phenomena. Thoughts come in from the external source and we almost think that they are our own, being unable to distinguish between our own thoughts and external

thoughts. The only time we really notice the externality would be when the external thoughts counteract our own and we feel "push back" on our own thoughts or actions. My guess is like watching a movie in your own body.

1. Passive external: We can sense or "hear" thoughts or feelings from other people and we can distinguish them from our own. They may have a different "voice" or what we could describe as "flavour", which is distinct from our own, but we cannot send thoughts back. You would be able to see everything going on somewhere else, but unable to do anything about it.

2. Active external: This probably represents the closest to what we would call "Shared Consciousness". We would be able to both send and receive information and be able to distinguish external information from internal.

These types of connections might enable people to at the simplest level pass basic information, at higher levels of bandwidth emotional feeds or visual information. This would require the self-model to be extended to include another person .This might start feeling like a super conscious being or hive mind. We are currently nowhere near getting the higher levels concepts operating and I expect that there would be significant physiological limitations. Put simply using magnets to generate signals in someone's brain could overheat their brain.

7.2.4 Transferring consciousness

Transferring consciousness requires that initially there are two separate entities that can become one and then are separated again. We have already discussed how sharing

So once the consciousness is separated again there would still be two entities each with independent memories, but very similar simulations. So (sadly) eternal life by cloning and transferring consciousness is not going to happen. At least we won't have eternal life this way.

Also if we were attempting to copy a human to an AI, I suspect we would find that the AI would not develop in the same way as the human. People learn, remember in a completely different way to computer systems. For example: An AI would have perfect memory recall, whereas a human does not. Learning, especially in deep networks, is also very different to humans, we need much less information to establish appropriate neural networks, so we might find that "AI" or AI copies, can recall everything they encounter, but learn slowly.

In more general terms the development of a consciousness seems to be heavily dependent on the hardware used to implement the consciousness.

7.3 Determinism and Free will

Research involving brain scans has shown that people have made a decision long before they realise it. Many researchers have suggested that this means we do not have free will. Also physicists have a problem with free will in that all of the calculations of physics show that outcomes are predictable based on the initial conditions of a system. This is referred to as Determinism.

Computer systems are deterministic. If we put in the same starting point of numbers we get exactly the same output. This is nicely shown in many of the previous examples considering fractals, cellular automata etc. Interestingly this might rule out the possibility that computers could become conscious at least according to Physics.

Physics can be is deterministic, but not always. If it is possible to determine the initial state of a physical system accurately enough the outcome of the system can be determined from a "few" "simple" rules. What this suggests is that if we could determine the initial state of the big bang we would be able to predict the rise of humanity, even the house you live in or the fact that you are reading this book right now could have been predicted based on the laws of physics. But I doubt it! One thing with physics is "Quantum foam" or the small scale structure of space-time, where "virtual particles" can spontaneously generate from nothing. Virtual particles are always pairs of particle and antiparticle that usually annihilate instantly. However, if they accidentally annihilated one of the virtual pair and one from the existing universe we still end up with a particle, but potentially with different velocity and momentum or even position. This, and a few other quantum physics mechanisms, would definitely change the parameters subtly of any real world system making them nondeterministic especially over large regions of space or time.

There is good news in this respect. From the model developed here it is possible to show that free will might exist even in computing. Humans have developed enough intelligence to "understand" the rules of physics. This allows us to use these rules to make predictions ourselves. When we find that our predictions are accurate we have a choice: Follow the predicted outcome or reject it and find another potential outcome if changes are made to the physical system. This is a paradox that might break the determinism of the universe.

This suggests that our consciousness, our ability to make conscious calculations based on their experience and understanding of the world gives a glimpse of the future. In the movie "Next" Nicholas Cage's character Cris Johnson says:

"Here is the thing about the future. Every time you look at, it changes, because you looked at it, and that changes everything else."

Conclusions – Answering some questions

His character is capable of seeing into the future, but never longer than a few minutes. This allows him to narrowly escape from people who are attempting to catch or kill him by looking at the possible outcomes of each decision. Whether or not you think it is a good movie, it demonstrates how knowledge of the can change its outcome.

Secondly, from the design, the simulation and the knowledge it creates are emergent. This means that even if you know exactly how the system works it does not allow you know the result of the simulation without running it.

In psychology there is a large body of work surrounding the subconscious and how this effects decisions. Many researchers in psychology, neurology and other scientific fields would agree that the subconscious has a significant effect on our consciousness. However, my feeling in this is that the subconscious can be described as "reflex decisions". Reflexes can be described as behaviours that are hard wired into a lifeform via hormones and other chemical effects in both the lifeform and the environment. For example life forms need food, hence most lifeforms will move towards food and eat it when they get hungry.

The next level of complexity in this concept it the idea of predators and prey. Let's think about two situations a snake vs an insect and a snake vs a mouse.

From what we understand of insects they have a very limited understanding of the world around them to the point that they typically will not have the sensor capabilities to detect a snake, as such it will not have an understanding of what a snake is or the ability to be conscious of a snake.

A mouse however does have an understanding of vision, smell and its environment. If the mouse sees or smells the snake it can make a conscious decision not to go near the snake, whereas the insect that does not have a conscious understanding of the snake will always go into an area that is occupied by the snake. Hence insects are more likely to be eaten by snakes.

The example of reflex vs conscious decisions is deliberately chosen here to show the difference between the outcomes. This is certainly an example of a difference in outcomes based on consciousness. If considering evolution the concept of consciousness in this example shows how consciousness <u>could</u> have evolved. The more resources a life form puts into its ability to simulate itself and its environment the more situations the lifeform could survive in and the more likely the lifeform is to survive in general.

From the brain scans mentioned previously it can be argued that because people have a lag between making a decision and realising that they have made a decision that this indicates people do not really have free will. Again it can be counter that the lag between the decision and realisation is simply the effect of the simulation itself and not a lack of free will. It is

Conclusions – Answering some questions

simply the process of calculating the outcomes of the situation. From a physics perspective we can still argue that even our conscious decisions can be influenced by what would be called initial conditions. My response to this argument is: this is true. If we know how people value the inputs of a decision it is possible to predict the outcome of the decision.

For example: a person is walking down a road and comes across a landmine. If they don't know what the landmine is or don't see it, well BOOM! If the person does "know" what the landmine is and "understands" what happens when a landmine is tripped they have a choice. The one most people would choose would be "Don't step on the mine". Note the key words "know" and "understand" that were defined in the introduction of this book. The knowledge of what landmines are comes from outside the person. The understanding comes from knowing how landmines work and the internal simulation of the situation. From the simulation we get the prediction that the landmine could kill anyone who gets too close. This example demonstrates the difference in outcomes from being able to predict the outcome of a situation.

I use a live or die example as this also demonstrates an evolutionary process. *{Secondly, it appeals to my morbid sense of humour.}* Life or death situations result in an evolutionary process, especially if we are dealing with capabilities that are transmittable to future generations. Clearly, humans have evolved effective simulations that allow accurate predictions. Hence we are able to survive many situations that would be difficult or impossible without such insights.

This relates to free will. There may be several possible outcomes for a particular situation. Depending on the inputs we may make a decision based purely on reflex which will result in a particular outcome. Alternatively it is possibly to consciously think about "the" situation and make a decision that has completely different outcomes to the "reflexive" decision.

Many researchers have noted that our unconsciousness runs the show most of the time. Consciousness seems to be required rarely, but it is active most of the time. So how much difference does consciousness need to make to be useful? My answer is: not much.

Even when completely conscious we may simply decide to follow our instinctive reactions. If they are well trained hopefully they are sensible decisions to make. My hypothesis is that consciousness is especially useful when we have conflicting reflexes. The important thing is that when we come across a new situation, a conscious decision can make a big difference in life. So while most of the time we can run as a deterministic zombie, the knowledge and simulation of the world allows us to make different decisions. So even if the brain is a deterministic simulation it allows the intelligence to make a prediction.

Conclusions – Answering some questions

This means that there is certainly a difference between reflex outcomes and a conscious decision. From this we can make the conclusion that a conscious decision is free will even if it is deterministic.

But the problem doesn't just end there. This also suggests that determinism and free will are not mutually exclusive. A counter argument is that if our intelligence is based on a deterministic system it is not really free will. So we could modify the statement to be "The illusion of free will" is not mutually exclusive with determinism. However I think we can say more. First, each person's simulation will have different understanding and knowledge as its basis, hence it is possible for different people to have different decisions. Secondly, we can take a person who learns new knowledge and see that they make different decisions. We can also look at what people "value" and see that these variations will have different outcomes. However, all this means is that people are predictable in many respects. For example, if you offer some-one a million dollars with no repercussions most people will take the money. This has nothing to do with determinism, it is simply that most people will make choices that will give themselves some advantage in life. We can only really test free will in situations where free will has an effect. Frankly most of the choices we make can be performed by reflexes or at least learned actions with no need for consciousness.

Even if the knowledge and simulation is deterministic, the variations of the knowledge base, understanding and personal values may not be deterministic in themselves, which adds another layer of complexity to the discussion. Especially when we consider "Values", we can consider many other factors and ideas that can be examined to test this idea.

The good news is that we can use the determinism vs free will arguments to justify a/the legal system. If we can assume that free will determines peoples actions based on an internal simulation which includes their knowledge and understanding of the world around them. We can consider that with what we decide are negative outcomes in the world why they occurred from the individuals perspective. In the case where the person did not realise that they caused a problem they require better awareness of the outside world. If they did not understand how their actions caused a problem, they need a better understanding of the consequences of our actions. In both cases they need a better simulation, but the solution is a better system of education, not necessarily a punishment. The last category is where a person deliberately performs an action that they have calculated from their simulation will cause a problem for someone else. We can split this up into two categories as well, by considering if their actions benefit themselves (some sort of zero sum situation), but still cause someone else a problem or if the actions are purely to cause harm to other people (e.g. Serial killers). In these cases (especially the last) they can be considered a criminal and probably need to be isolated from the rest of the community. Even given this analysis I think the legal system is far more interested in punishing people than considering education as a solution in many cases... but that is another bigger discussion for another time. So I will summarise by stating that having a theory that confirms free will also has legal implications.

I am sure that many researchers can add far more evidence and arguments to this discussion. But for brevity, I will simply reiterate that it <u>seems</u> the ability to make a choice based on a simulation that produces a different outcome from reflexes or our existing subconscious process, indicates that free will exists.

If researchers can accept the idea that conscious decisions indicate free will, this idea can provide a basis to determine what creatures are conscious and to a certain extent to what degree.

7.4   Is it a Cartesian Theatre?

Sort of. The Cartesian theatre (Coined by Dennet) refers to is the idea of Plato's allegory of the cave, where people watch light and shadows of the world outside the cave projected onto the cave wall. They have no experience of the real world but they can infer the content of the real world from the images on the world. The idea with the Cartesian theatre is that we experience consciousness as people ("homunculi") watching the light show.

The design takes on this idea in that we have many independent sensory analysis tools, "object" recognisers. This idea is similar to the ideas of Baars [19], [67], [2], [150], Graziano [151] and perhaps Dennet [14], [15]. The main difference in the analogy is that I would put the observers experiencing the external world. Each observer has very specific abilities to observe features of the world. A few of them, the ones in working memory, draw their specific interpretations of the world on the cave wall and see if it matches up with the shadows and lights that fall on the wall.

In this way the Cartesian theatre exists, but what we produce in terms of internal imagers or sensory experience is what we are actually conscious of.

7.5   Is this design really consciousness?

Short Version: ... I don't know.

Conclusions – Answering some questions

This section goes through a few question that at present I can't answer. The design itself is based around the system analysis of sensory data running through our brains with the output being the actions we perform, much of which is aimed at keeping us alive. My feeling is that this gives a good description of how information is generated, processed and understood by our brains.

But is it actually conscious? Short version: I don't know.

But I think it also gives a better model than previous ones found in other fields. If anything is missing, this model should allow us to find a better explanations for any gap in the theory.

Is there anything left?

If the consciousness loop does not itself create consciousness is there room for something that could be internal to the "Consciousness Loop" that would allow a quantum effect or some other processing for consciousness to emerge. My feeling is that we can tell we are conscious. Demonstrating how the brain can achieve this is another matter.

Penrose could be right in that quantum fluctuations in cellular microtubules could have an underlying basis for consciousness. But, we still don't know how these fluctuations would cause consciousness to emerge. My feeling is that if this was the case it would be related to the consciousness loop central to the design. Again I have no idea of how much of a gap remains between the information processing concepts and whatever else a quantum or other spiritual ideas bring to the system.

Would this design be able to pass a Turing test? ... With training ... possibly. But not necessarily. "The Design" allows for intelligence to be low, so the entity has a limited understanding of the world but can still be conscious. Conversely, it might be possible for a highly intelligent computer system to imitate a human being, but not be conscious. I would argue that this is the case with existing AI's Watson, Alpha Go etc. They are highly capable of finding answers to questions or making tactical decisions to outsmart an opponent, but as far as we know they are not conscious. These implementations show that there are still huge differences between a computer implementation and a life form. A computer can shut down completely, a biological life form can't. Computers could have perfect recall, Biologicals do not. Does this make a difference?

So ultimately does this design answer the question of what consciousness is?

The only way to be sure of this would be to "be" the entity this is implemented in. However, from an external point of view there would be no way to tell. So this discussion might be a moot point.

Hence: Shrug!

7.6 Consciousness Redefined

From the discussions in previous chapters we have established the design as a sensible interpretation of what occurs in the mind to make consciousness. We can redefine some of our terms using the design as a basis for new definitions. Feel free to disagree with the definitions generated, the point is to show that these new definitions are based off the design itself.

7.6.1 Perception

The ability to perceive objects becomes the ability to match a template generated from a simulation to incoming sensory data.

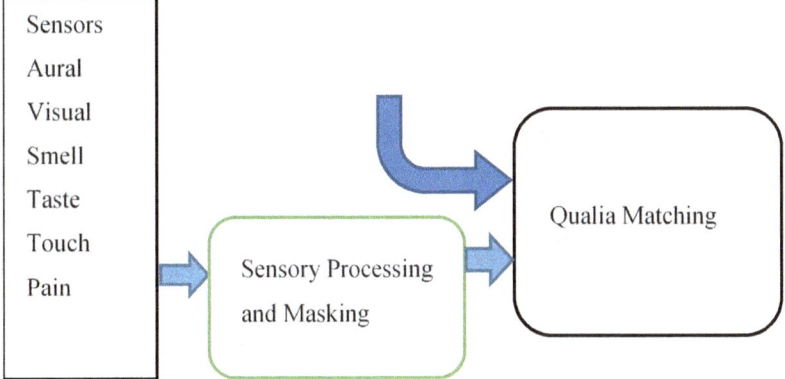

Figure 111 Design of Consciousness simplified

7.6.2 Awareness

From the design, we can consider that awareness is the process of generating a template and matching it so incoming sensory stimulus. This creates a template in the simulation or working memory. Thus, we can consider awareness to be the object in working memory being matched with sensory stimulus which includes an extra step from the design.

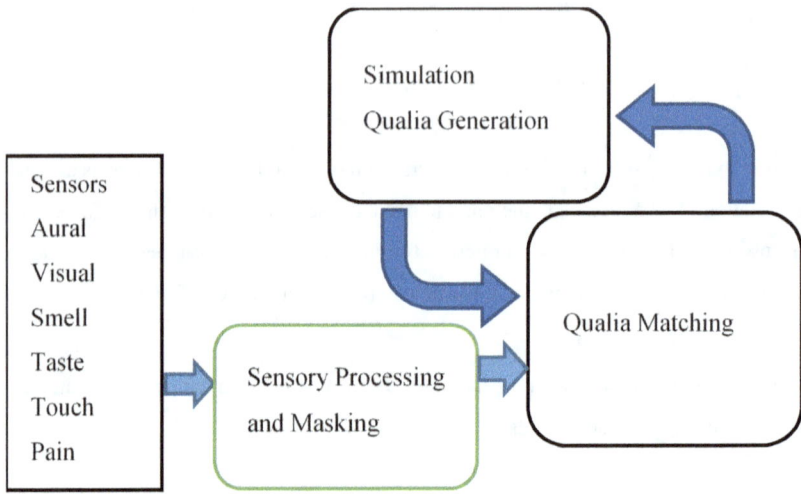

Figure 112 Design of Consciousness simplified

Awareness can also be thought of as a loop where the sensory information matched to the simulation data is analysed and matches to the templates that are used in the simulation.

This suggests that to be aware of something means that the object is a stable part of the data running through the consciousness loop.

7.6.3 Understanding

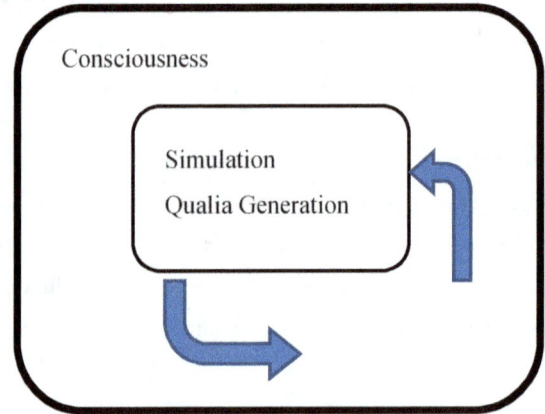

Figure 113 Understanding from the Design of Consciousness simplified

Understanding can be thought of as the ability to simulate an objects actions. The simulation represents knowledge about how an object changes in the world. It is active knowledge as opposed to static "knowledge".

Conclusions – Answering some questions

7.6.4 *Consciousness*

From the discussion presented in this book we can create a better definition of what consciousness is:

Consciousness is a state of "awareness" of both external stimulus and reinforced by internal qualia. This state works closely with memories, allowing them to be archived and recalled. Given knowledge and a functional understanding of a situation, consciousness allows predictions to be made about the situation. This gives a choice to follow the expected outcome or to attempt to change the situation. This also allows new behaviours to be generated and old behaviours to be actively examined and updated.

A simpler version would be:

Consciousness is the internal simulation of ourselves and the world around ourselves.

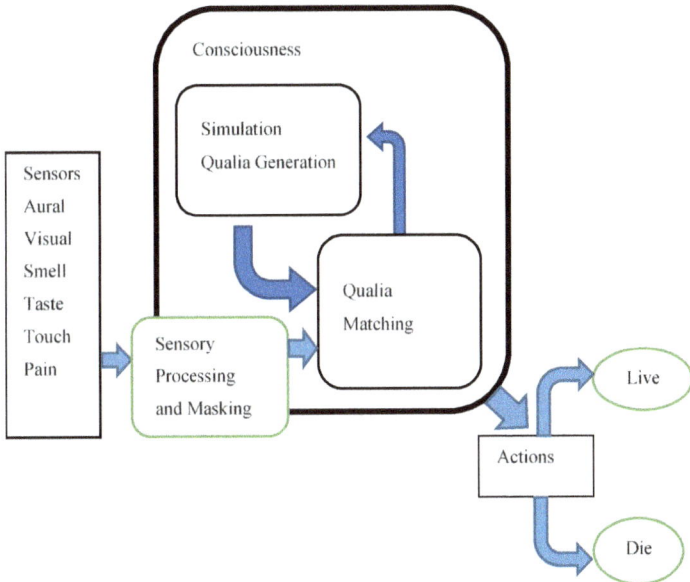

Figure 114 Design of Consciousness simplified

Conclusions – Answering some questions

8 Finally

Even if this idea isn't correct it is still a worthwhile conversation to have. What has been developed in this book can be used as a basis for testing ideas about consciousness further. So even if the design is wrong it allows us to consider how to test for consciousness, so feel free to modify or add to this design and continue the discussion.

I showed that the features of a conscious algorithm would include:

1. Data
2. Operations
3. Feedback loops
4. Generative system
5. Predictive

The question is: Does this get us beyond the limitations of algorithms to something that can breach the mind body boundary?

My feeling is yes... but ultimately, I leave it to you.

Finally

9 Bibliography

[1] B. J. Baars, "In the theatre of consciousness. Global workspace theory, a rigorous scientific theory of consciousness," *J. Conscious. Stud.*, vol. 4, no. 4, pp. 292–309, 1997.

[2] B. J. Baars, "Global workspace theory of consciousness: toward a cognitive neuroscience of human experience," *Prog. Brain Res.*, vol. 150, pp. 45–53, 2005.

[3] A. K. Seth, B. J. Baars, and D. B. Edelman, "Criteria for consciousness in humans and other mammals," *Conscious. Cogn.*, vol. 14, no. 1, pp. 119–139, 2005.

[4] A. K. Seth and B. J. Baars, "Neural Darwinism and consciousness," *Conscious. Cogn.*, vol. 14, no. 1, pp. 140–168, 2005.

[5] J. Gleick, "Chaos: Making a New Science," *N. Y. Viking*, 1987.

[6] J. Conway, "Conway's Game of Life." http://pi.math.cornell.edu/~lipa/mec/lesson6.html

[7] I. Kassabalidis, M. El-Sharkawi, R. Marks, P. Arabshahi, and A. Gray, "Swarm intelligence for routing in communication networks," in *GLOBECOM'01. IEEE Global Telecommunications Conference (Cat. No. 01CH37270)*, 2001, vol. 6, pp. 3613–3617.

[8] G. Rzevski and P. Skobelev, "Emergent intelligence in large scale multi-agent systems," *Int. J. Educ. Inf. Technol.*, vol. 1, no. 2, pp. 64–71, 2007.

[9] H. Iba, *AI and SWARM: Evolutionary Approach to Emergent Intelligence*. CRC Press, 2019.

[10] J. R. Searle, D. C. Dennett, and D. J. Chalmers, *The mystery of consciousness*. New York Review of Books, 1997.

[11] C. Francis, "The Astonishing Hypothesis, The Scientific Search for the Soul," *N. Y. Charles Scribner'S Sons*, 1994.

[12] R. Penrose, "The emperor's new mind," *RSA J.*, vol. 139, no. 5420, pp. 506–514, 1991.

[13] S. Pinker, "How the mind works," *Ann. N. Y. Acad. Sci.*, vol. 882, no. 1, pp. 119–127, 1999.

[14] D. C. Dennett, *From bacteria to Bach and back: The evolution of minds*. WW Norton & Company, 2017.

[15] D. C. Dennett, *Consciousness explained*. Penguin uk, 1993.

[16] R. Penrose, *Shadows of the Mind*, vol. 4. Oxford University Press Oxford, 1994.

[17] C. Koch, *The Feeling of Life Itself: Why Consciousness is Widespread But Can't be Computed*. Mit Press, 2019.

[18] C. Koch, M. Massimini, M. Boly, and G. Tononi, "Neural correlates of consciousness: progress and problems," *Nat. Rev. Neurosci.*, vol. 17, no. 5, p. 307, 2016.

[19] Barnard J Baars, *On Consciousness: Science & Subjectivity - Updated Works on Global Workspace Theory*. New York, USA: Nautilus Press, 2019.

[20] P. Carruthers, *The centered mind: What the science of working memory shows us about the nature of human thought*. OUP Oxford, 2015.

[21] G. M. Edelman, *The remembered present: a biological theory of consciousness*. Basic Books, 1989.

[22] C. Fernyhough, *The voices within: The history and science of how we talk to ourselves*. Basic Books, 2016.

[23] D. Eagleman, *The brain: The story of you*. Canongate Books, 2015.

Bibliography

[24] G. M. Edelman, *Bright air, brilliant fire: On the matter of the mind*. Basic books, 1992.

[25] G. M. Edelman, "Neural Darwinism: selection and reentrant signaling in higher brain function," *Neuron*, vol. 10, no. 2, pp. 115–125, 1993.

[26] G. Edelman, "Consciousness: The Remembered Present," *Ann. N. Y. Acad. Sci.*, vol. 929, no. 1, pp. 111–122, 2001, doi: 10.1111/j.1749-6632.2001.tb05711.x.

[27] G. M. Edelman, J. A. Gally, and B. J. Baars, "Biology of consciousness," *Front. Psychol.*, vol. 2, p. 4, 2011.

[28] E. Kocagoncu, A. Klimovich-Gray, L. E. Hughes, and J. B. Rowe, "Evidence and implications of abnormal predictive coding in dementia," *ArXiv Prepr. ArXiv200606311*, 2020.

[29] A. Witon *et al.*, "Sedation modulates fronto-temporal predictive coding circuits and the double surprise acceleration effect," *Cereb. Cortex*, 2020.

[30] S. L. Denham and I. Winkler, "Predictive coding in auditory perception: challenges and unresolved questions," *Eur. J. Neurosci.*, vol. 51, no. 5, pp. 1151–1160, 2020.

[31] P. Kok and F. P. de Lange, "Predictive coding in sensory cortex," in *An introduction to model-based cognitive neuroscience*, Springer, 2015, pp. 221–244.

[32] M. Strauss *et al.*, "Disruption of hierarchical predictive coding during sleep," *Proc. Natl. Acad. Sci.*, p. 201501026, 2015.

[33] W. H. Alexander and J. W. Brown, "Frontal cortex function as derived from hierarchical predictive coding," *Sci. Rep.*, vol. 8, no. 1, p. 3843, Mar. 2018, doi: 10.1038/s41598-018-21407-9.

[34] Z. Straka, T. Svoboda, and M. Hoffmann, "PreCNet: Next Frame Video Prediction Based on Predictive Coding," *ArXiv Prepr. ArXiv200414878*, 2020.

[35] M. W. Spratling, "A Hierarchical Predictive Coding Model of Object Recognition in Natural Images," *Cogn. Comput.*, vol. 9, no. 2, pp. 151–167, Apr. 2017, doi: 10.1007/s12559-016-9445-1.

[36] Y. Huang and R. P. Rao, "Predictive coding," *Wiley Interdiscip. Rev. Cogn. Sci.*, vol. 2, no. 5, pp. 580–593, 2011.

[37] B. Millidge, "Combining active inference and hierarchical predictive coding: A tutorial introduction and case study," 2019.

[38] J. Pöppel and S. Kopp, "Satisficing models of bayesian theory of mind for explaining behavior of differently uncertain agents: Socially interactive agents track," in *Proceedings of the 17th international conference on autonomous agents and multiagent systems*, 2018, pp. 470–478.

[39] K. J. Friston, W. Wiese, and J. A. Hobson, "Sentience and the origins of consciousness: From Cartesian duality to Markovian monism," *Entropy*, vol. 22, no. 5, p. 516, 2020.

[40] J. Hohwy and A. Seth, "Predictive processing as a systematic basis for identifying the neural correlates of consciousness," *Philos. Mind Sci.*, vol. 1, no. II, 2020.

[41] T. Parr, G. Rees, and K. J. Friston, "Computational neuropsychology and Bayesian inference," *Front. Hum. Neurosci.*, vol. 12, p. 61, 2018.

[42] K. J. Friston, T. Parr, and B. de Vries, "The graphical brain: belief propagation and active inference," *Netw. Neurosci.*, vol. 1, no. 4, pp. 381–414, 2017.

[43] N. Rabinowitz, F. Perbet, F. Song, C. Zhang, S. A. Eslami, and M. Botvinick, "Machine theory of mind," in *International conference on machine learning*, 2018, pp. 4218–4227.

Bibliography

[44] C. Baker, R. Saxe, and J. Tenenbaum, "Bayesian theory of mind: Modeling joint belief-desire attribution," in *Proceedings of the annual meeting of the cognitive science society*, 2011, vol. 33, no. 33.

[45] S. Freud, "On Psychotherapy (1905 [1904])," in *The Standard Edition of the Complete Psychological Works of Sigmund Freud, Volume VII (1901-1905): A Case of Hysteria, Three Essays on Sexuality and Other Works*, 1953, pp. 255–268.

[46] S. Freud, "The Unconscious," 1915.

[47] S. Freud, "A General Introduction to Psychoanalysis (24th edn)," 1924.

[48] S. Freud, "The ego and the id. The standard edition of the complete psychological works of Sigmund Freud. vol. 19," *Lond. Hogarth*, 1923.

[49] S. Freud, *Complete psychological works of Sigmund Freud*, vol. 14. Random House, 2001.

[50] M. Lewis, "Self-knowledge and social development in early life," *Handb. Personal. Theory Res.*, pp. 277–300, 1990.

[51] R. F. Baumeister, *The Self*. Oxford university press, 2010.

[52] I. Goodfellow *et al.*, "Generative adversarial nets," in *Advances in neural information processing systems*, 2014, pp. 2672–2680.

[53] I. J. Goodfellow, "On distinguishability criteria for estimating generative models," *ArXiv Prepr. ArXiv14126515*, 2014.

[54] I. J. Goodfellow *et al.*, "Generative adversarial networks," *ArXiv Prepr. ArXiv14062661*, vol. 4, no. 5, p. 6, 2014.

[55] I. Goodfellow, "NIPS 2016 tutorial: Generative adversarial networks," *ArXiv Prepr. ArXiv170100160*, 2016.

[56] T. Gonsalves, "Board games ai," in *Advanced Methodologies and Technologies in Artificial Intelligence, Computer Simulation, and Human-Computer Interaction*, IGI Global, 2019, pp. 68–80.

[57] F. Li and Y. Du, "From AlphaGo to power system AI: What engineers can learn from solving the most complex board game," *IEEE Power Energy Mag.*, vol. 16, no. 2, pp. 76–84, 2018.

[58] S. D. Holcomb, W. K. Porter, S. V. Ault, G. Mao, and J. Wang, "Overview on deepmind and its alphago zero ai," in *Proceedings of the 2018 international conference on big data and education*, 2018, pp. 67–71.

[59] J.-F. Yeh, P.-H. Su, S.-H. Huang, and T.-C. Chiang, "Snake game AI: Movement rating functions and evolutionary algorithm-based optimization," in *2016 Conference on Technologies and Applications of Artificial Intelligence (TAAI)*, 2016, pp. 256–261.

[60] S. Bhatti, A. Desmaison, O. Miksik, N. Nardelli, N. Siddharth, and P. H. Torr, "Playing doom with slam-augmented deep reinforcement learning," *ArXiv Prepr. ArXiv161200380*, 2016.

[61] M. Wydmuch, M. Kempka, and W. Jaśkowski, "Vizdoom competitions: Playing doom from pixels," *IEEE Trans. Games*, vol. 11, no. 3, pp. 248–259, 2018.

[62] A. Diamond, R. Knight, D. Devereux, and O. Holland, "Anthropomimetic Robots: Concept, Construction and Modelling," *Int. J. Adv. Robot. Syst.*, vol. 9, no. 5, p. 209, 2012, doi: 10.5772/52421.

[63] O. Holland, "Can a Virtual Entity Support Real Consciousness, and How Might This Lead to Conscious Robots?," in *AAAI Spring Symposium: Towards Conscious AI Systems*, 2019.

[64] A. M. Turing, "COMPUTING MACHINERY AND INTELLIGENCE," *Mind*, vol. 59, no. 236, pp. 433–460, 1950.

Bibliography

[65] M. W. Spratling, "A Hierarchical Predictive Coding Model of Object Recognition in Natural Images," *Cogn. Comput.*, vol. 9, no. 2, pp. 151–167, 2017, doi: 10.1007/s12559-016-9445-1.

[66] B. J. Baars, W. P. Banks, and J. B. Newman, *Essential sources in the scientific study of consciousness*. Mit Press, 2003.

[67] G. M. Edelman, J. A. Gally, and B. J. Baars, "Biology of consciousness," *Front. Psychol.*, vol. 2, p. 4, 2011.

[68] F. Crick, C. Koch, G. Kreiman, and I. Fried, "Consciousness and neurosurgery," *Neurosurgery*, vol. 55, no. 2, pp. 273–282, 2004.

[69] D. J. Chalmers, "Consciousness and Cognition," *AI CD-ROM Revis.*, vol. 2, 1990.

[70] N. Wolpe, F. H. Hezemans, and J. B. Rowe, "Alien limb syndrome: A Bayesian account of unwanted actions," *Cortex*, 2020, [Online]. Available: https://www.sciencedirect.com/science/article/pii/S0010945220300538

[71] F. de A. A. Gondim, J. W. L. T. Júnior, A. A. Morais, P. M. G. Sales, and H. G. Wagner, "Alien limb syndrome responsive to amantadine in a patient with corticobasal syndrome," *Tremor Hyperkinetic Mov.*, vol. 5, 2015, [Online]. Available: https://www.ncbi.nlm.nih.gov/pmc/articles/PMC4505076/

[72] J. Graff-Radford *et al.*, "The alien limb phenomenon," *J. Neurol.*, vol. 260, no. 7, pp. 1880–1888, 2013.

[73] D. Tiwari and K. Amar, "A case of corticobasal degeneration presenting with alien limb syndrome," *Age Ageing*, vol. 37, no. 5, pp. 600–601, 2008.

[74] O. Rabbani, L. E. Bowen, R. T. Watson, E. Valenstein, and M. S. Okun, "Alien limb syndrome and moya–moya disease," *Mov. Disord. Off. J. Mov. Disord. Soc.*, vol. 19, no. 11, pp. 1317–1320, 2004.

[75] K. A. Josephs and M. N. Rossor, "The alien limb," *Pract. Neurol.*, vol. 4, no. 1, pp. 44–45, 2004.

[76] A. R. Mayes and N. Roberts, "Theories of episodic memory," *Philos. Trans. R. Soc. Lond. B. Biol. Sci.*, vol. 356, no. 1413, pp. 1395–1408, 2001.

[77] H. J. Markowitsch, A. Thiel, M. Reinkemeier, J. Kessler, A. Koyuncu, and H. Wolf-Dieter, "Right amygdalar and temporofrontal activation during autobiographic, but not during fictitious memory retrieval.," *Behav. Neurol.*, vol. 12, no. 4, pp. 181–190, 2000, doi: 10.1155/2000/303651.

[78] W. L. Chafe, "Language and consciousness," *Language*, pp. 111–133, 1974.

[79] J. R. Searle, S. Willis, and others, *Consciousness and language*. Cambridge University Press, 2002.

[80] J. Zlatev, "The dependence of language on consciousness," *J. Conscious. Stud.*, vol. 15, no. 6, pp. 34–62, 2008.

[81] T. D. Thao, *Investigations into the origin of language and consciousness*, vol. 44. Springer Science & Business Media, 2012.

[82] T. Parr, A. W. Corcoran, K. J. Friston, and J. Hohwy, "Perceptual awareness and active inference," *Neurosci. Conscious.*, vol. 2019, no. 1, 2019, doi: 10.1093/nc/niz012.

[83] John Horgan, "Are Brains Bayesian?," *Scientific American*, Jan. 06, 2016. https://blogs.scientificamerican.com/cross-check/are-brains-bayesian/ (accessed Mar. 29, 2021).

[84] D. W. Smith, "Phenomenology," *Stanford Encyclopedia of Philosophy*, 2013. https://plato.stanford.edu/entries/phenomenology/ (accessed Mar. 14, 2021).

[85] Josh Weisberg, "The Hard Problem of Consciousness," *Internet Encyclopedia of Philosphy*, 2013. https://iep.utm.edu/hard-con/#SH3b

Bibliography

[86] Janet Levin, "Functionalism," *Stanford Encyclopedia of Philosophy*, 2018. https://plato.stanford.edu/entries/functionalism/ (accessed Mar. 14, 2021).

[87] H. Von Helmholtz, *Handbuch der physiologischen Optik: mit 213 in den Text eingedruckten Holzschnitten und 11 Tafeln*, vol. 9. Voss, 1867.

[88] Raúl Arrabales Moreno, A. S. de Miguel, and A. L. Espino, "Evaluation and development of consciousness in artificial cognitive systems," *Ph Diss.*, 2011.

[89] R. Arrabales, A. Ledezma, and A. Sanchis, "ConsScale: A pragmatic scale for measuring the level of consciousness in artificial agents," *J. Conscious. Stud.*, vol. 17, no. 3–4, pp. 131–164, 2010.

[90] L. Ross, D. Greene, and P. House, "The 'false consensus effect': An egocentric bias in social perception and attribution processes," *J. Exp. Soc. Psychol.*, vol. 13, no. 3, pp. 279–301, 1977, doi: https://doi.org/10.1016/0022-1031(77)90049-X.

[91] J. K. Oostrom, N. C. Köbis, R. Ronay, and M. Cremers, "False consensus in situational judgment tests: What would others do?," *J. Res. Personal.*, vol. 71, pp. 33–45, 2017.

[92] M. D. Coleman, "Emotion and the false consensus effect," *Curr. Psychol.*, vol. 37, no. 1, pp. 58–64, 2018.

[93] D. Wolpert, K. Pearson, C. Ghez, and E. Kandel, "Principles of Neural Science," *Organ. Plan. Mov. 5th Ed N. Y. McGraw-Hill*, pp. 475–97, 2013.

[94] S. Dehaene, *Consciousness and the brain: Deciphering how the brain codes our thoughts*. Penguin, 2014.

[95] J. Goodman and M. Packard, "Memory systems of the basal ganglia," in *Handbook of Behavioral Neuroscience*, vol. 24, Elsevier, 2016, pp. 725–740.

[96] F. McNab and T. Klingberg, "Prefrontal cortex and basal ganglia control access to working memory," *Nat. Neurosci.*, vol. 11, no. 1, pp. 103–107, 2008.

[97] M. G. Packard and B. J. Knowlton, "Learning and memory functions of the basal ganglia," *Annu. Rev. Neurosci.*, vol. 25, no. 1, pp. 563–593, 2002.

[98] A. M. Graybiel, "Building action repertoires: memory and learning functions of the basal ganglia," *Curr. Opin. Neurobiol.*, vol. 5, no. 6, pp. 733–741, 1995.

[99] J. LeDoux, "The amygdala," *Curr. Biol.*, vol. 17, no. 20, pp. R868–R874, 2007.

[100] R. V. Bretas, M. Taoka, H. Suzuki, and A. Iriki, "Secondary somatosensory cortex of primates: beyond body maps, toward conscious self-in-the-world maps," *Exp. Brain Res.*, vol. 238, no. 2, pp. 259–272, Feb. 2020, doi: 10.1007/s00221-020-05727-9.

[101] M. D. Greicius, B. Krasnow, A. L. Reiss, and V. Menon, "Functional connectivity in the resting brain: a network analysis of the default mode hypothesis," *Proc. Natl. Acad. Sci.*, vol. 100, no. 1, pp. 253–258, 2003.

[102] A. Horn, D. Ostwald, M. Reisert, and F. Blankenburg, "The structural–functional connectome and the default mode network of the human brain," *Neuroimage*, vol. 102, pp. 142–151, 2014.

[103] M. E. Raichle, "The Brain's Default Mode Network," *Annu. Rev. Neurosci.*, vol. 38, no. 1, pp. 433–447, 2015, doi: 10.1146/annurev-neuro-071013-014030.

[104] C. F. Gillespie, S. T. Szabo, and C. B. Nemeroff, "Chapter 36 - Unipolar depression," in *Rosenberg's Molecular and Genetic Basis of Neurological and Psychiatric Disease (Sixth Edition)*, Sixth Edition., R. N. Rosenberg and J. M. Pascual, Eds. Academic Press, 2020, pp. 613–631. doi: 10.1016/B978-0-12-813866-3.00036-9.

Bibliography

[105] S. Kumar, S. Joseph, P. E. Gander, N. Barascud, A. R. Halpern, and T. D. Griffiths, "A Brain System for Auditory Working Memory," *J. Neurosci.*, vol. 36, no. 16, pp. 4492–4505, 2016, doi: 10.1523/JNEUROSCI.4341-14.2016.

[106] N. Müller and R. Knight, "The functional neuroanatomy of working memory: contributions of human brain lesion studies," *Neuroscience*, vol. 139, no. 1, pp. 51–58, 2006.

[107] E. Salmon *et al.*, "Regional brain activity during working memory tasks," *Brain*, vol. 119, no. 5, pp. 1617–1625, 1996, doi: 10.1093/brain/119.5.1617.

[108] J. Eriksson, E. K. Vogel, A. Lansner, F. Bergström, and L. Nyberg, "Neurocognitive Architecture of Working Memory," *Neuron*, vol. 88, no. 1, pp. 33–46, Oct. 2015, doi: 10.1016/j.neuron.2015.09.020.

[109] W. J. Chai, A. I. Abd Hamid, and J. M. Abdullah, "Working Memory From the Psychological and Neurosciences Perspectives: A Review," *Front. Psychol.*, vol. 9, p. 401, 2018, doi: 10.3389/fpsyg.2018.00401.

[110] T. Parr and K. J. Friston, "Working memory, attention, and salience in active inference," *Sci. Rep.*, vol. 7, no. 1, pp. 1–21, 2017.

[111] S. Grossberg, "The link between brain learning, attention, and consciousness," *Conscious. Cogn.*, vol. 8, no. 1, pp. 1–44, 1999.

[112] J. A. Hobson and E. F. Pace-Schott, "The cognitive neuroscience of sleep: neuronal systems, consciousness and learning," *Nat. Rev. Neurosci.*, vol. 3, no. 9, pp. 679–693, 2002.

[113] D. R. Bevan, "Curare: from laboratory to law court," *Clin. Invest. Med.*, vol. 41, pp. 17–20, 2018.

[114] H. R. GRIFFITH, "CURARE IN ANESTHESIA," *J. Am. Med. Assoc.*, vol. 127, no. 11, pp. 642–644, 1945, doi: 10.1001/jama.1945.02860110022006.

[115] N. S. White and M. T. Alkire, "Impaired thalamocortical connectivity in humans during general-anesthetic-induced unconsciousness," *Neuroimage*, vol. 19, no. 2, pp. 402–411, 2003.

[116] A. G. Hudetz, "General anesthesia and human brain connectivity," *Brain Connect.*, vol. 2, no. 6, pp. 291–302, 2012.

[117] A. G. Hudetz and G. A. Mashour, "Disconnecting Consciousness: Is There a Common Anesthetic End Point?," *Anesth. Analg.*, vol. 123, no. 5, pp. 1228–1240, Nov. 2016, doi: 10.1213/ANE.0000000000001353.

[118] M. J. Redinbaugh *et al.*, "Thalamus modulates consciousness via layer-specific control of cortex," *Neuron*, 2020.

[119] J. Pardey, S. Roberts, L. Tarassenko, and J. Stradling, "A new approach to the analysis of the human sleep/wakefulness continuum," *J. Sleep Res.*, vol. 5, no. 4, pp. 201–210, 1996.

[120] A. Mack, I. Rock, and others, *Inattentional blindness*. MIT press, 1998.

[121] L. Dotto, "Sleep stages, memory and learning.," *CMAJ Can. Med. Assoc. J.*, vol. 154, no. 8, p. 1193, 1996.

[122] M. P. Walker and R. Stickgold, "Sleep-dependent learning and memory consolidation," *Neuron*, vol. 44, no. 1, pp. 121–133, 2004.

[123] L. Mascetti, A. Foret, M. Bonjean, L. Matarazzo, T. Dang-Vu, and P. Maquet, "Some facts about sleep relevant for Landau-Kleffner syndrome," *Epilepsia*, vol. 50, pp. 43–46, 2009.

[124] E. A. McDevitt, K. A. Duggan, and S. C. Mednick, "REM sleep rescues learning from interference," *Neurobiol. Learn. Mem.*, vol. 122, pp. 51–62, 2015.

Bibliography

[125] J. W. Astington and M. J. Edward, "The development of theory of mind in early childhood," *Encycl. Early Child. Dev.*, pp. 1–6, 2010.

[126] H. M. Wellman, *The child's theory of mind*. The MIT Press, 1992.

[127] G. Akdeniz, S. Toker, and I. Atli, "Neural mechanisms underlying visual pareidolia processing: An fMRI study," *Pak. J. Med. Sci.*, vol. 34, no. 6, p. 1560, 2018.

[128] Anil Seth, "Models of consciousness," *http://www.scholarpedia.org*, 2007. http://www.scholarpedia.org/article/Models_of_consciousness

[129] G. M. Edelman, *The remembered present: a biological theory of consciousness*. Basic Books, 1989.

[130] O. Blanke, T. Landis, L. Spinelli, and M. Seeck, "Out-of-body experience and autoscopy of neurological origin," *Brain*, vol. 127, no. 2, pp. 243–258, 2004.

[131] O. Blanke and S. Arzy, "The out-of-body experience: disturbed self-processing at the temporo-parietal junction," *The Neuroscientist*, vol. 11, no. 1, pp. 16–24, 2005.

[132] O. Blanke et al., "Linking out-of-body experience and self processing to mental own-body imagery at the temporoparietal junction," *J. Neurosci.*, vol. 25, no. 3, pp. 550–557, 2005.

[133] J. Aspell and O. Blanke, "Understanding the out-of-body experience from a neuroscientific perspective," Nova Science Publishers, 2009.

[134] H.-D. Park and O. Blanke, "Coupling inner and outer body for self-consciousness," *Trends Cogn. Sci.*, vol. 23, no. 5, pp. 377–388, 2019.

[135] L. Tófoli and D. de Araujo, "Treating addiction: Perspectives from EEG and imaging studies on psychedelics," in *International review of neurobiology*, vol. 129, Elsevier, 2016, pp. 157–185.

[136] R. L. Carhart-Harris et al., "Neural correlates of the psychedelic state as determined by fMRI studies with psilocybin," *Proc. Natl. Acad. Sci.*, vol. 109, no. 6, pp. 2138–2143, 2012.

[137] C. G. Davey, J. Pujol, and B. J. Harrison, "Mapping the self in the brain's default mode network," *Neuroimage*, vol. 132, pp. 390–397, 2016.

[138] R. Carhart-Harris, R. Leech, E. Tagliazucchi, and others, "How do hallucinogens work on the brain," *J. Psychophysiol.*, vol. 71, no. 1, pp. 2–8, 2014.

[139] R. L. Carhart-Harris et al., "Neural correlates of the LSD experience revealed by multimodal neuroimaging," *Proc. Natl. Acad. Sci.*, vol. 113, no. 17, pp. 4853–4858, 2016.

[140] D. J. Chalmers, "What is a neural correlate of consciousness," *Neural Correl. Conscious. Empir. Concept. Quest.*, pp. 17–39, 2000.

[141] E. S. Yudkowsky, "The AI-Box Experiment," 2002. https://www.yudkowsky.net/singularity/aibox

[142] E. Yudkowsky and others, "Artificial intelligence as a positive and negative factor in global risk," *Glob. Catastrophic Risks*, vol. 1, no. 303, p. 184, 2008.

[143] E. Yudkowsky, N. Bostrom, and M. Cirkovic, "Global catastrophic risks," *N Bostrom MM Cirković N. Y. Oxf. Univ. Press Chap Cogn. Biases Potential. Affect. Judgm. Glob. Risks*, 2008.

[144] N. Bostrom and E. Yudkowsky, "The ethics of artificial intelligence," *Camb. Handb. Artif. Intell.*, vol. 1, pp. 316–334, 2014.

[145] E. Yudkowsky, *Inadequate Equilibria: Where and How Civilizations Get Stuck*. Machine Intelligence Research Institute, 2017.

Bibliography

[146] T. Collins and M. Prigg, "Facebook shuts down controversial chatbot experiment after AIs develop their own language to talk to each other," *Abgerufen Unter Httpwww Dailymail Co Uksciencetecharticle-4747914Facebook-Shuts-Chatbotsmake-Lang. Html Stand Vom 04-04-2018*, 2017.

[147] R. Kucera, "The truth behind Facebook AI inventing a new language," *Data Sci.*, vol. 7, p. 2017, 2017.

[148] A. Stocco *et al.*, "Playing 20 questions with the mind: collaborative problem solving by humans using a brain-to-brain interface," *PloS One*, vol. 10, no. 9, p. e0137303, 2015.

[149] R. P. N. Rao *et al.*, "A direct brain-to-brain interface in humans," *PloS One*, vol. 9, no. 11, pp. e111332–e111332, Nov. 2014, doi: 10.1371/journal.pone.0111332.

[150] B. J. Baars, "Global workspace theory of consciousness: toward a cognitive neuroscience of human experience," *Prog. Brain Res.*, vol. 150, pp. 45–53, 2005.

[151] M. S. Graziano, "Consciousness engineered," *J. Conscious. Stud.*, vol. 23, no. 11–12, pp. 98–115, 2016.

Bibliography

{

Quick while he's not looking do something funny!

Like What?

I don't know, do a fart joke.

This is an academic level book on consciousness and you want to do a fart joke?

Sure, Irony.

That's a bit lame. We could explore, the meaning of life, the meaning of consciousness, some aspect of consciousness split personalities, zombie consciousness.

Zombie "Braaaaaaaaaiiiiins"

Oh Great you woke up the Zombie.

Zombie killer: Kill it, Kill it, Kill it, Kill iiit

Odd that seems to come up a lot.

[Thwack Thwack Thwack Thwack Thwack Thwack]

They make a lot of mess.

Gamer guy: BOOM! Headshot!

Spray pattern analyst: Looks pretty though.

Robot Steve: That does not compute.

Pirate Steve: Aaaargh! Make him walk the plank!

Physics guy: Planks constant: $6.62607004 \times 10^{-34}$

Maybe we should just go back to building robots

Evil genius: Excellent My Evil Robot Shall rule the World! Muhuhuhuhuhuhuhuhuhuhahahahahahahah!!!

}

Will you lot shut up!

The end.

www.ingramcontent.com/pod-product-compliance
Lightning Source LLC
Chambersburg PA
CBHW070249010526
44107CB00056B/2395